EARLY
EMBRYOLOGY
OF THE
CHICK

EARLY EMBRYOLOGY OF THE CHICK

BRADLEY M. PATTEN

Emeritus Professor of Anatomy
University of Michigan Medical School

Fifth Edition

McGraw-Hill Book Company

New York   St. Louis   San Francisco   Düsseldorf   Johannesburg
Kuala Lumpur   London   Mexico   Montreal   New Delhi   Panama
Rio de Janeiro   Singapore   Sydney   Toronto

EARLY EMBRYOLOGY OF THE CHICK

*This book was set in Patina by University Graphics, Inc., and
printed on permanent paper by Halliday Lithograph Corporation,
and bound by the Book Press, Inc. The designer was Paula Tuerk.
The inserts were printed by Lehigh Press Lithographers. The editors
were Jeremy Robinson and Janet Wagner. Sally Ellyson supervised
production.*

CONTENTS

PREFACE

The fact that chick embryos are so generally used as laboratory material for courses in vertebrate embryology warrants the treatment of their development in a book designed primarily for the beginning student. To one commencing the study of embryology the very abundance of the information available is likely to be confusing and discouraging. He has difficulty in culling the essentials and fitting them together in their proper relationships and tends to become lost in a maze of detail. This book has been written in an effort to set forth for him in brief and simple form the basic mechanisms of embryology. It does not purport to be a reference work. Details and controverted points have been avoided for the sake of clarity in outlining fundamental processes.

The story of development as illustrated by the chick has been taken as the basic scheme of presentation. This departure from the conventional comparative method of approach does not imply any undervaluation of the comparative point of view. On the contrary, I am convinced that it is a goal to be sought by anyone who would consider himself trained in embryology; and, wherever it seemed possible without breaking the thread of the story of chick development, processes have been interpreted from this broader outlook. But for the beginner to attempt to follow the development of several forms simultaneously seems to me irrational. A student new to the subject is merely confused when confronted at the outset by a mass of comparative data. What a beginner needs is not a vast array of facts to be memorized, but the thread of a coherent story to hold together the new facts he acquires. The story of the development of a single form, told without digression, has an inherent quality of sustained interest, which is of inestimable value in creating an understanding of embryological processes. Building on such a foundation, each new excursion into the broader field of comparative embryology becomes more stimulating because its findings are progressively more significant.

It has seemed inexpedient, likewise, to confront beginners with an extensive discussion of experimental embryology, fascinating as is this approach to the study of development. The proper interpretation of embryological experiments depends on first acquiring a basic knowledge of the normal course of development. It is such a foundation that this book aims to present. Moreover, the techniques employed and the laboratory equipment necessary make experimental work much more suitable for advanced courses with small numbers of students than for introductory courses. I have, therefore, limited myself in this field to including a few examples of experiments which seemed especially helpful in interpreting the underlying mechanisms of normal development.

Because of my conviction that so-called elementary texts too frequently overreach their avowed scope, the ground covered by this book has been rigidly restricted. The account of development has been carried only through the first four days of incubation, the later developmental processes being dismissed with mere statements about the adult structures that are derived from the various organs of the embryo. The reasons for thus devoting the major part of the book to the early phases of development would seem quite obvious. In this period the body of the embryo is laid down and the organ systems are well established. To one at all familiar with the adult structure of vertebrates it is relatively easy to understand the later changes in the position and proportions of organs if he has seen how they are first formed. Furthermore, courses in elementary embryology rarely continue work on the chick beyond the early stages of its development, and more extensive courses, in which a knowledge of mammalian embryology is the objective, ordinarily pass from the study of three- or four-day chicks to work on mammalian embryos.

Although the text has been kept brief, illustrations have been freely used in the belief that they convey ideas more readily than can be done in writing. Much time and care has been used in an effort to make each figure convey its message with clearness and accuracy. The increased size of the pages in the present edition as compared with earlier editions has permitted making the new illustrations of a generous size and has made possible the enlarging of a number of the original figures which had suffered by too great reduction. These changes should materially improve the usability of the book as a whole.

In revising the text, advantage was taken of the great volume of important embryological research that has appeared in recent years. However, neither the alterations in the text nor the additional illustrations have changed the essential character of the book. It remains as it was originally planned, a simply presented story of the early embryology of the chick.

BRADLEY M. PATTEN
Ann Arbor, Michigan
December 1969

ACKNOWLEDGMENTS

In the shaping of the first edition of this book I was most deeply indebted to my father, William Patten, late Professor of Biology and Evolution at Dartmouth College. From his broad teaching experience he gave me constructive suggestions about planning the text and especially about making the illustrations. In the preparation of the whole-mounts and serial sections on which the work was based, I had the able assistance of Mrs. Mary V. Bayes, then embryological technician in the Department of Histology and Embryology at Western Reserve University School of Medicine.

The revisions for the fourth and fifth editions were made after I had left Western Reserve to go to the University of Michigan. Many of my colleagues there have generously aided in these more recent phases of the work. Dr. Alexander Barry has been particularly helpful in his critical reading of the text. Miss Cecelia Banwell gave me highly skilled assistance with some of the new drawings, and Dr. Theodore C. Kramer was a resourceful collaborator whenever matters of projection or photography were involved.

In the preparation of all the editions I received generous assistance from my wife in converting my deviously interlined longhand manuscript into usable typescript. Moreover, she had an unerring ability to put her fingers on passages that were not quite clearly expressed, so they could be revised before they found their way into print.

The first four editions of the book were published by the Blakiston Company. Throughout the work they were most generously cooperative. The unusually faithful reproductions of the original drawings, especially those in color, were due to the understanding and expert supervision of Mr. Willard T. Shoener. In this new edition the same efficient and thoughtful cooperation has been given by the College Division of the McGraw-Hill Book Company.

BRADLEY M. PATTEN

EARLY
EMBRYOLOGY
OF THE
CHICK

CHAPTER 1

INTRODUCTION

Embryology   Every one of the higher animals starts life as a single cell—the fertilized ovum. This fertilized ovum, as its technical name *zygote* implies, has a dual origin. It is formed by the fusion of a germ cell from the male parent with one from the female parent. The union of two such sex cells to form a zygote constitutes the process of *fertilization* and initiates the life of a new individual. Embryology is the study of the growth and differentiation undergone by an organism in the course of its development from a single fertilized egg cell into a highly complex and independent living being like its parents.

As a study embryology offers more than the mere acquisition of an interesting array of facts. It gives one an understanding of some of the ways of life. We are all egoists, to a certain extent at least, and anything that touches the matter of our own whence and whither is of absorbing interest. The processes by which a fish, or an alligator, or a chick grows from a single fertilized egg cell to its fully elaborated adult structure are fundamentally the same as those involved in our own development (Fig. 1). And these growth processes hold for us something definite and tangible in answer to that ever recurring question, "Whence do we come, and how?"

Embryology is also an important source of evidence as to the path fol-

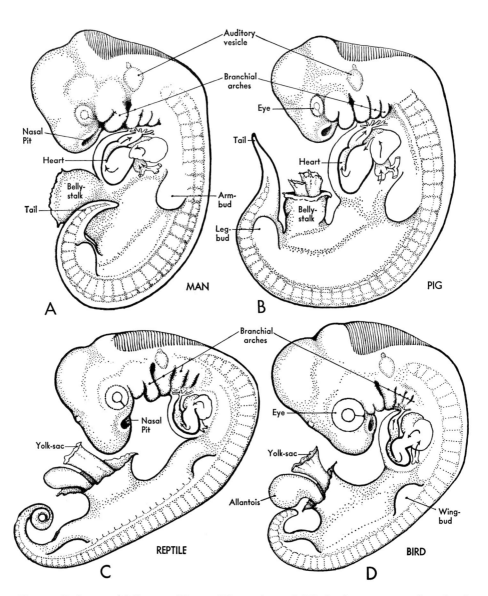

Fig. 1. Embryos of (A) man, (B) pig, (C) reptile, and (D), bird at corresponding developmental stages. The striking resemblance of the embryos to one another is indicative of the fundamental similarity of the processes involved in their development. (*From William Patten, "Evolution," Dartmouth College Press, Hanover, N.H., 1922.*)

lowed by evolution. It tells us in one short, uninterrupted story how each individual grows into an adult. We can see this process going on under our very eyes. And we know that the story of individual development sketches for us in outline the evolutionary changes of our forbears. For the law of biogenesis or recapitulation is that *every living thing, in its individual development, passes through a series of constructive stages like those in the evolutionary development of the race to which it belongs.*

> This means that there is but one main way to upbuild a given kind of organism, and that every individual must do it in essentially the same way his ancestors did. But to-day the individual can do this much more quickly and economically than its ancestors, because it does not have to find out by long and costly experimenting just how to do it. Essentially the right ways and means of doing it "automatically" are provided for it by its antecedents. At every step in the process it is using something which Nature, out of all the efforts of the past, has provided for it. These ways and means constitute the Heritages of individual life, and range all the way from ancestral germinal materials, stored up foods, and other parental provisions, to the environmental physical and social provisions in which the growing organism is placed.[1]

More tangible perhaps than the perspective embryology gives on the life of today and on its evolutionary history in the past, is its direct help in the study of anatomy. Not until the student becomes enmeshed in a maze of structural details does he realize his imperative need of a knowledge of how and why adult conditions became as they are. For only this knowledge will lead him beyond blind memorizing to comprehension. And when he delves still deeper into anatomy, he begins to encounter puzzling variations from the normal body architecture. Sometimes these are merely minor anomalies which do not materially affect the functional fitness of the individual; sometimes they are extensive malformations which render continued life precarious or even altogether impossible. Our understanding of such conditions, and what future hope there may be of reducing the frequency with which they occur, can come only through extending our knowledge of genetics and embryology.

The only method of attaining a comprehensive understanding of embryological processes is through the study and comparison of development in various animals. Many phases of the development of any specific organism can be interpreted only through a knowledge of corresponding processes

[1] From William Patten: "Evolution," part II, The Evolution of Life, Dartmouth College Press, Hanover, N.H., 1922.

in other organisms. The beginning student, however, can most readily and with least risk of confusion acquire his knowledge of embryology through intensive study of one form at a time. Building on the familiarity with fundamental processes of development thus acquired, he may later broaden his horizon by the comparative study of a variety of forms.

The Chick as Laboratory Material The chick is one of the most satisfactory animals on which student laboratory work in embryology may be based. Chick embryos in a proper state of preservation and of the stages desired can readily be secured and prepared for study. Used as the only laboratory material in a brief course, they afford a basis for understanding the early differentiation of the organ systems and the fundamental processes of body formation common to all groups of vertebrates. In more extended courses where several forms are taken up, the chick serves at once as a type for the development characteristic of the large-yolked eggs of birds and reptiles and as an intermediate form bridging the gap between the simpler processes of development in fishes and amphibia on the one hand and the more complex processes in mammals on the other. In medical school courses where a knowledge of human embryology is the end in view, the chick not only makes a good stepping stone to the understanding of mammalian embryology, but also provides material for the study of early developmental processes not readily demonstrable in human material.

Plan and Scope of This Book This book on the development of the chick has been written for those who are beginning the study of embryology and has accordingly been kept as brief and uncomplicated as possible. Nevertheless it is assumed that the beginner in embryology will not be without a certain background of zoological knowledge and training. He may reasonably be expected to be familiar with the fundamental facts of evolution and heredity, the structure of cells and their methods of division, the nature of the various types of tissues, and the more general phases of the morphology of vertebrates. It therefore seems unnecessary to include here any preliminary discussion of these phenomena.

Because I am convinced that it is not a logical approach for a beginner, this book does not emphasize the comparative viewpoint. Nevertheless, where the comparative background for some particular phase of chick development seemed to promise to make it more intelligible, I have not hesitated to bring it in. This has been done especially freely in connection with cleavage and gastrulation in the early parts of the book, and in connection with the cardiovascular and excretory systems in the final chapter. It is hoped that these excursions, instead of rendering the student's task more onerous, will lighten it by making these phases of development more readily understandable.

It has seemed wise to take essentially the same standpoint with reference to experimental embryology. This rapidly growing and important field is one which can be handled best in an advanced course, building on a background such as this book aims to provide for beginners. Accordingly, no attempt has been made to deal with it systematically. But, as with comparative embryology, when the experimental approach seemed to offer particular help in interpreting some phase of development, it has been freely drawn upon. References for collateral reading on these and other phases of the subject are given in the bibliography.

Methods of Study　　Like other sciences embryology demands first of all accurate observation. It differs considerably, however, from such a science as adult anatomy where the objects studied are relatively constant and their component parts are not subject to rapid changes in their interrelations. During development, structural conditions within the embryo are constantly changing. Each phase of development presents a new complex of conditions and new problems.

Solution of the problems presented in any given stage of development depends upon a knowledge of the stages which precede it. To comprehend the embryology of an organism one must, therefore, start at the beginning of its development and follow in their natural order the changes which occur. At the outset of his work the student must realize that proper sequence of study is essential and may not be disregarded. A knowledge of structural conditions in earlier stages than that at the moment under consideration and an appreciation of the trend of the developmental processes by which conditions at one stage become transmuted into different conditions in the next are direct and necessary factors in acquiring a real comprehension of the subject. Without them the story of embryology becomes incoherent, a mere jumble of confused impressions.

A knowledge of the phenomena of development is ordinarily acquired by studying a series of embryos at various stages of advancement. Each stage should be studied not so much for itself, as for the evidence it affords of the progress of development. In the study of embryology it does not suffice to acquire merely a series of "still pictures" of various structures, however accurate these pictures may be. The study demands a constant application of correlative reasoning and an appreciation of the mechanical factors involved in the relations to each other of various structures within the embryo, and in the relation of the embryo as a whole to its environment. In order really to comprehend the embryologic significance of a structure one must know not only its relations within the embryo being studied at the time, but also the manner in which it has been derived and the nature of the changes by which it is progressing toward adult conditions. To get absolutely the whole story it is obvious that one would have to study a series of embryos with infinitely small intervals between them. Nevertheless the fundamental steps in the process may be grasped from a much less extensive series. The fewer the stages studied, however, the more careful must one be to keep in mind the continuity of the processes and to think out the changes by which one stage leads to the next.

The outstanding idea to be kept in mind by the student beginning the study of embryology is that the development of an individual is a process and that this process is continuous. The conditions he sees in embryos of various stages are of importance chiefly because they serve as evidence of events in the process of development at various intervals in its continuity, as historical events are evidence of the progress of a nation. Just as historical events are led up to by preparatory occurrences and followed by results which in turn affect later events, so in embryology events in development are presaged by preliminary changes and when consummated affect in turn later steps in the process.

In certain respects the laboratory study of embryological material involves methods of work for which courses in general zoology do not entirely prepare the student. Some general suggestions as to methods of procedure are, therefore, not out of place.

In dissecting gross material it is not unduly difficult to appreciate the complete relationships of a structure. The nature of embryological material, however, introduces new problems. Embryos of the age when the establishment of the various organ systems and processes of body formation are being initiated are too small to admit of successful dissection, but not sufficiently small to permit of the satisfactory microscopical study of an intact embryo, except for its more general organization. To study embryos of this stage with any degree of thoroughness, they must be cut into sections which are sufficiently thin to allow effective use of the microscope to ascertain cellular organization and detailed structural relationships. In preparing such material the entire embryo is cut into sections which are mounted on slides in the order in which they were cut. A sectional view of any region of the embryo is then available for study.

While sections readily yield accurate information about local regions, it is extremely difficult to construct a mental picture of any whole organism from a study of serial sections alone. For this reason it is necessary to work first on entire embryos which have been prepared by staining and clearing so they may be studied as transparent objects. From such preparations it is possible to map out the configuration of the body and the location and extent of the more conspicuous internal organs. In this work the fact that embryos have three dimensions must be kept constantly in mind, and, by careful focusing, the depth at which a structure lies must be determined as well as its apparent position in surface view. Although, conventionally, entire chick embryos are usually represented in dorsal view, much additional information may be gained by following a study of the dorsal, with a study of the ventral aspect. Unless the preliminary study of entire embryos is carefully and thoughtfully carried out, the study of sections is likely to yield only confusion.

In studying a section from a series it is necessary first of all to determine the location in which it was cut through the embryo. The plane of the section under consideration, as well as the region of the embryo through which it passes, should be ascertained by comparing it with an entire embryo of the same age as that from which the section was cut. Only when the location of a section is known precisely can the structures appearing in it be correlated with the organization of the embryo as a whole. Probably nothing in the study of embryology causes students more difficulties than neglect to locate sections accurately, with the consequent failure to appreciate the relationships of the structures seen in them. Too great emphasis cannot be laid on the vital importance of fitting the structures shown by sections properly into the general scheme of organization as it appears in whole-mounts (Figs. 2 and 3).

It must by no means be inferred that the possibilities of the whole-mounts have been exhausted by the preliminary study accorded them before taking up the work on sections. Further and more careful study of entire embryos should constantly accompany the study of serial sections. Many details which in the initial observation of the whole-mount were inconspicuous or abstruse will become significant in the light of the more exact information yielded by the sections. Relative levels of closely related structures which are particularly difficult to work out by focusing through a whole-mount are unequivocally shown by sections. Conversely, there is nothing that will help a beginner as much in the interpretation of serial sections as constantly relating them to the organization of the entire embryo. Correlative study of the whole-mount and the sections is, veritably, the key to success in the study of young embryos.

In embryology it is necessary to designate the location of structures and

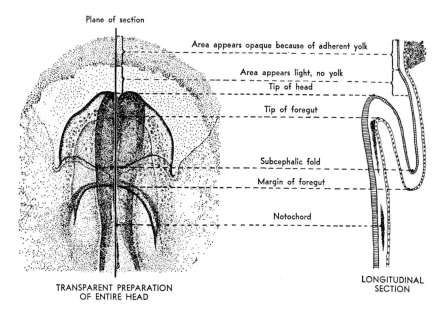

Plane of section

Area appears opaque because of adherent yolk

Area appears light, no yolk
Tip of head

Tip of foregut

Subcephalic fold

Margin of foregut

Notochord

LONGITUDINAL
SECTION

TRANSPARENT PREPARATION
OF ENTIRE HEAD

Fig. 2. Relation of longitudinal section of the embryonic head to the picture presented by a head of the same age mounted entire as a transparent preparation.

the direction of growth processes by terms which are referable to the body of the embryo regardless of the position it occupies. Our ordinary terms of location which are referred to the direction of the action of gravity, such as above, over, under, etc., are not sufficiently accurate because of the fact that the embryo itself may lie in a great variety of positions.[1] The correct adjectives of position are dorsal, pertaining to the back; ventral, pertaining to the belly; cephalic, to the head; caudal, to the tail; mesial, to the middle part; and lateral, to the side of the embryonic body. In dealing with relations within the head the adjective cephalic is inadequate and the term rostral (Latin, beak of a bird; bow of a ship) is used to designate the extreme anterior portion of the head, or intra-cephalic structures such as the brain. Adverbs of fixed position are made in the usual way be adding "-ly" to the root of the adjective.

In addition to adverbs of position, corresponding adverbs of motion or

[1] In gross human anatomy there still persist many terms that are referred to gravity. Such terms, because of the erect posture of man, are not applicable to comparative anatomy or to embryology. The most confusing of these are anterior and posterior as used to mean, respectively, pertaining to the belly and to the back. In comparative anatomy and in embryology, anterior has reference to the head region and posterior to the tail region. Because of this possible source of confusion, the terms anterior and posterior should be replaced by their more precise synonyms, cephalic and caudal, wherever the context is such that their significance might otherwise be doubtful.

direction are formed by adding the suffix "-ad" to the root of the adjective, as dorsad, meaning toward the back, cephalad, meaning toward the head, etc. These adverbs should be applied only to the progress of processes, or to the extension of structures toward the part indicated by their root. Thus, for example, we should say that the developing eye of an embryo was lo-

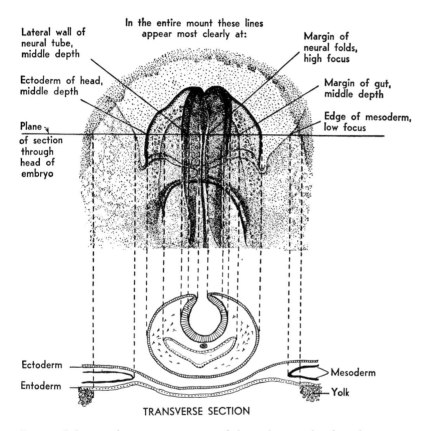

Fig. 3. Relation of transverse section of the embryonic head to the picture presented by an entire head of the same age viewed as a transparent preparation. The two drawings may be brought into closer relation by looking at them with the top of the page tilted downward. This and the preceding figure show the aid that sections can be in interpreting entire embryos and, at the same time, the way in which the greater perspective lent by study of entire embryos increases the significance of sectional pictures. Note especially that neither the transverse nor the longitudinal section alone adequately interprets all the features of the entire mount. An embryo is a three-dimensional object and must be studied with that fact in mind. These figures show, also, how largely the appearance of lines or bands seen in studying transparent mounts of entire embryos is due to looking edgewise through a folded layer.

cated in the lateral wall of the forebrain or that the forebrain was in the cephalic part of the embryo; but if we wished to express the idea that as the eye increased in size it moved farther to the side we should say it grew laterad. Cultivation of the use of correct and definite terms of position and direction in dealing with embryological processes will greatly aid accurate thinking and clear understanding.

CHAPTER 2

THE GAMETES AND FERTILIZATION

Continuity of Germ Plasm   The reproductive cells which unite to initiate the development of a new individual are known as *gametes*. In the case of all the higher organisms the gametes are of two types produced by two sexually differentiated individuals, the small, actively motile gametes from the male being called *spermatozoa* or *spermia*, and the larger, food-laden gametes formed within the female being termed *ova*. The gametes and the cells which give rise to them are said to constitute the individual's *germ plasm*. The cells which take no direct part in the production of gametes are called somatic cells. In antithesis to the germ plasm, the somatic cells collectively are said to constitute the *somatoplasm*.

The germ plasm is of paramount interest, not only to the biologist, but to all thinking persons, because while the somatic cells cease to exist with the death of the individual whose body they constitute, the germ plasm may live on indefinitely in succeeding generations. In the higher vertebrates where sexual reproduction is the rule, the germ plasm of a single individual cannot survive by itself. There must be successful union of a male with a female sex cell. These two conjugating gametes alone pass on the entire hereditary dowry of the species. It is not easy to realize fully all that is implied by this simple statement as to the continuity of the germ plasm. It may make it more vivid if we apply it specifically to ourselves and say that the future of the human race depends on the germ plasm held in trust within the bodies of the individuals now living. Fortunately, very early in the life of an individual, the germ plasm is segregated in the gonads and is not subject to most of the vicissitudes and the diseases from which the somatic cells suffer. But the germ plasm, even though it is not directly affected, may nevertheless suffer indirectly because of a poor environment forced upon it by an unhealthy body.

The quality of the germ plasm of any one individual, while of great importance, is only half the story. Of equal significance is the nature of the combination of germ plasm which occurs in each generation when the male and female gametes fuse. As surely as either gamete brings into the new combination defective germ plasm, so surely will both the body and the germ plasm of the new individual suffer therefrom.

History of Sex Cells within the Parent Body   It is, therefore, of fundamental interest to go back of the production of gametes in the sexually mature individual so we may see all the steps in the preservation and transmission of the racial heritage. Obviously the germ plasm of any individual must have come to it from its parents by way of the ovum and the spermium. But one wants to know how and where the germ plasm was cared for by the parents; when in their life history it first became possible to recognize it as distinct from the somatoplasm; when and how it became segregated in the gonads; and what it was doing during the long period before sexual maturity.

There is still much of the very early history of the germ plasm which is but imperfectly known. Yet the cells which are destined to give rise to the gametes are definitely recognizable at a surprisingly early stage in development. Even before it is possible to tell whether an embryo is destined to become a male or a female, certain large cells, different from their neighbors, can be recognized as the *primordial sex cells*. In other words we know that they constitute the germ plasm as it exists in an embryo of that age.

The primordial sex cells are first easily identifiable when the gonads are just taking shape as definite organs. Until recently it was believed that, in the higher vertebrates, they could not be recognized as potential germ cells any earlier in their history. Recently much more detailed investigations have been made which seem to indicate that the sex cells can be recognized even prior to their appearance in the newly formed gonad. According to these studies the primordial germ cells become recognizably differentiated in the wall of the yolk-sac, and migrate thence to become established in the growing gonads (Fig. 4). How much farther back toward the fertilized ovum the lineage of the primordial sex cells may in the future be traced, it would be unwise to predict. Even now, in some of the invertebrates, it is believed that the single cell which gives rise to all the sex cells can be identified when the embryo is so young that it is no more than a minute ball of cells.

The upper part of the chart appearing as Fig. 5 summarizes graphically the conception of the early separation, from the somatic cells, of certain cells which are destined to give rise to the gametes. Since this process occurs before sexual differentiation, we may take it as a common starting point for the germ-cell lineage of either sex. The lower part of the chart outlines separately for each sex the later history of the germ cells. Sexual differentiation of the embryo begins to be apparent shortly after the primordial germ cells are es-

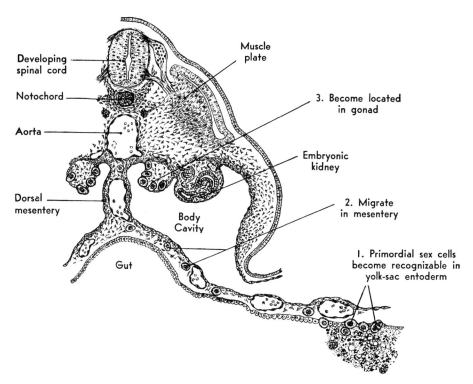

Fig. 4. Schematic section through the midbody region of a young embryo, illustrating the manner in which the primordial germ cells are believed to originate in the yolk-sac entoderm and migrate thence to the developing gonad.

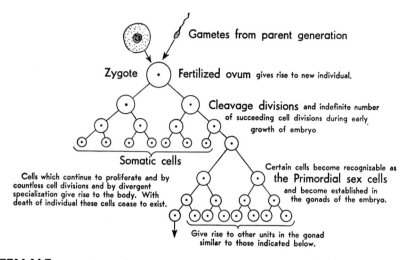

Gametes from parent generation

Zygote • Fertilized ovum gives rise to new individual.

Cleavage divisions and indefinite number of succeeding cell divisions during early growth of embryo

Somatic cells

Cells which continue to proliferate and by countless cell divisions and by divergent specialization give rise to the body. With death of individual these cells cease to exist.

Certain cells become recognizable as the Primordial sex cells and become established in the gonads of the embryo.

Give rise to other units in the gonad similar to those indicated below.

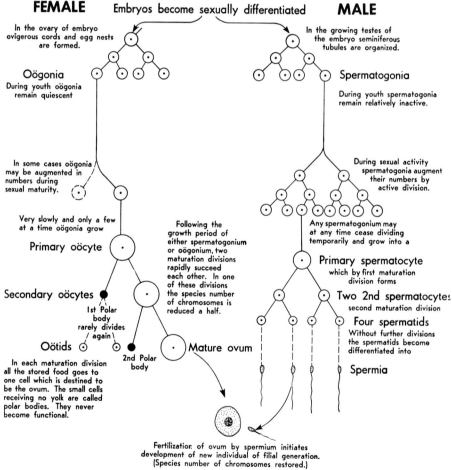

FEMALE   Embryos become sexually differentiated   **MALE**

In the ovary of embryo ovigerous cords and egg nests are formed.

In the growing testes of the embryo seminiferous tubules are organized.

Oögonia
During youth oögonia remain quiescent

Spermatogonia

During youth spermatogonia remain relatively inactive.

In some cases oögonia may be augmented in numbers during sexual maturity.

During sexual activity spermatogonia augment their numbers by active division.

Very slowly and only a few at a time oögonia grow

Following the growth period of either spermatogonium or oögonium, two maturation divisions rapidly succeed each other. In one of these divisions the species number of chromosomes is reduced a half.

Any spermatogonium may at any time cease dividing temporarily and grow into a

Primary oöcyte

Primary spermatocyte
which by first maturation division forms

Secondary oöcytes
1st Polar body rarely divides again

Two 2nd spermatocytes
second maturation division

Four spermatids
Without further divisions the spermatids become differentiated into

Oötids
2nd Polar body

Mature ovum

In each maturation division all the stored food goes to one cell which is destined to be the ovum. The small cells receiving no yolk are called polar bodies. They never become functional.

Spermia

Fertilization of ovum by spermium initiates development of new individual of filial generation. (Species number of chromosomes restored.)

Fig. 5.   Chart outlining, for one generation, the history of the gametes and the germ plasm from which they are derived.

tablished in the gonads. If the individual is to become a male, the gonads differentiate into testes. During early embryonic life the primordial sex cells become organized within the testis in the form of tortuous tubules called the *seminiferous tubules*. During the period of body growth which precedes sexual maturity, the future gamete-producing cells or *spermatogonia* in these tubules remain quiescent. During the period of active sexual life, the spermatogonia multiply and groups of them are constantly undergoing the final changes which precede their liberation as mature gametes (Fig. 5).

The changes undergone by the female sex cells are, in a general way, comparable to those passed through by the male cells. The large primordial cells of the germ plasm form cords and clusters in the embryonic ovary, homologous with the seminiferous tubules in the testis. The unmatured egg-mother cells comparable to the spermatogonia of the male are called *oögonia* (Fig. 5). Not all the oögonia are destined to become functional gametes. Many of them remain dormant, while only relatively few receive sufficient nutrition to progress toward maturity. The selection of certain oögonia for special nutrition is an early expression of the most characteristic difference between the male and female gametes. The male gametes contain no stored food material and are small and active. There is no special nutritional provision for any favored few. But such enormous numbers of them are produced that under normal conditions there is little chance that an elaborately prepared female gamete will go to waste because it is not found and fertilized by a spermium. The development of large-yolked ova is on a quite different basis. Of all the potential gametes contained in the ovary, relatively few are destined to be brought to maturity. But these few are richly endowed with stored food materials. The energy expended by the male on quantity production of small active gametes ensuring fertilization is, thus, in the female, devoted to laying up a supply of food which provides the necessary materials for the growth of the embryo which is to result from fertilization.

Spermatogenesis   The latter part of the history of the sex cells which occurs just before their liberation as mature gametes demands closer scrutiny. In the male these terminal changes are spoken of as constituting the process of spermatogenesis. The cells of the germ line which came to be located in the seminiferous tubules of the growing testis we have already learned to know by their technical name of spermatogonia (Fig. 5). When a male animal becomes sexually mature these spermatogonia begin to produce gametes. They do not all become active at any one time, but periodically groups of spermatogonia begin to produce successive crops of mature spermia.

The process is much the same in all the higher vertebrates and it will

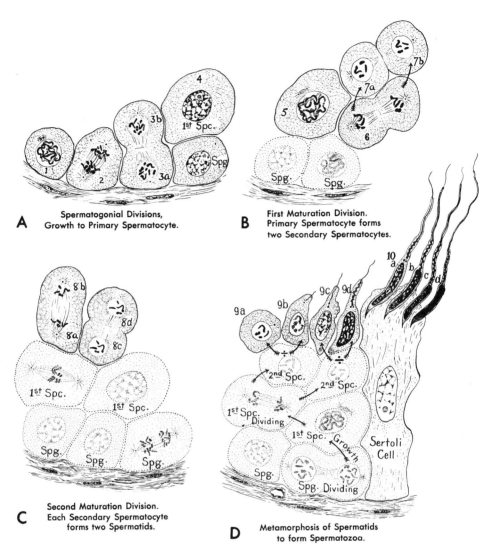

A   Spermatogonial Divisions, Growth to Primary Spermatocyte.

B   First Maturation Division. Primary Spermatocyte forms two Secondary Spermatocytes.

C   Second Maturation Division. Each Secondary Spermatocyte forms two Spermatids.

D   Metamorphosis of Spermatids to form Spermatozoa.

Fig. 6. Semischematic diagrams to show the main steps in the progress of spermatogenesis. In the wall of a mature seminiferous tubule, cells in all stages of spermatogenesis occur in such close association that it is not always easy to grasp their relations. To obviate this difficulty, these diagrams start with conditions in a young tubule just going into activity and follow the process a step at a time. The sequence of events is further indicated by consecutive numbering of the cells. In each part of the figure, cells undergoing the critical changes there emphasized are drawn in full detail, while cells not at the moment under consideration are "shadowed in" to make evident the accompanying changes in positional relations. For the sake of simplicity the species number of chromosomes is assumed to be eight.

Abbreviations: 1st Spc., primary spermatocyte; *2nd Spc.,* secondary spermatocyte; *Spg.,* spermatogonium.

be at once simpler and more profitable if we trace it in broad general lines rather than attempting to master its details in any one form. In sections of the testis, spermatogonia are found peripherally located in the walls of the seminiferous tubules. When a spermatogonium undergoes mitosis [Fig. 6, A (1–3)], either of the resulting cells may do one of two things. It may cease dividing for a time, and by growing to a size markedly larger than that of its parent, become differentiated as a *primary spermatocyte* [Fig. 6, A(4)]. Or it may remain like its parent [Fig. 6, A(3a and Spg.)] and take the place of cells which crowd toward the lumen of the tubule [Fig. 6, A(3b)] to become differentiated into spermatocytes. Some of the spermatogonial cells always remain thus in the peripheral part of the tubule and furnish a constant source of new cells ready for conversion into spermatocytes.

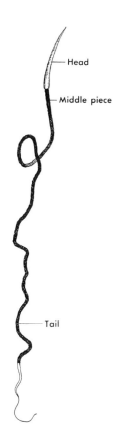

Head

Middle piece

Tail

Fig. 7. Spermatozoon of the pigeon. (*After Ballowitz.*)

Once a cell has passed through the growth phase which so differentiates it that it is called a primary spermatocyte, its future history is definitely determined. As soon as its growth is completed, it undergoes two divisions in rapid succession. The first of these divisions results in the formation of two smaller cells called *secondary spermatocytes* [Fig. 6, B(7a, 7b)]. Each of these secondary spermatocytes, without any resting period which might allow the cells to grow to the size attained by their parents, promptly divides again. The four small cells thus produced from the primary spermatocyte by two successive cell divisions are known as *spermatids* [Fig. 6, C(8a, b, c, d)]. The two divisions are known as the *maturation (meiotic) divisions* both because they are the last divisons that these cells undergo and because they accomplish certain important internal changes preparatory to the part these cells are to play in fertilization. The nature of these changes, especially as they affect the chromosomal content of the cells, we shall consider in more detail in connection with the similar changes taking place in the maturation of the female sex cells.

Although cell division ceases with the two maturation divisions, the spermatids still have radical changes to undergo before they are capable of carrying out their function. During this *metamorphosis of the spermatid* the nuclear material becomes very compact to form the bulk of the head of the spermium. The flagellum-like *tail* is formed by the development of a contractile fibril which first makes its appearance in the cytoplasm near the centrosome and then rapidly grows in length until it projects far beyond the original confines of the cell. As this tail filament grows it remains enveloped in a thin film of cytoplasm which eventually forms a membrane covering all but the extreme tip of the tail [Fig. 6, *D*(9a–d, 10)]. The modified centrosomal apparatus becomes located in the so-called "neck region" where the "tail" (flagellum) is joined to the "head" of the spermium. The cytoplasm of the spermatid is reduced greatly in bulk forming an almost invisibly thin envelope about the nuclear material of the head. At the extreme tip of the head a cap-like thickening of modified cytoplasm constitutes the acrosome. Thus a mature male gamete is a cell consisting essentially of a very compact nucleus provided with a flagellum which gives it the power of active locomotion in a fluid medium (Fig. 7).

Oögenesis   Spermatogenesis could be dealt with in general terms because it is fundamentally the same in the entire vertebrate group. The processes involved in the formation of ova, although they also show a certain underlying similarity in all the great groups of vertebrates, are greatly modified by differences in the amount of food material stored as yolk in the egg cells. The amount of yolk characteristically present in the egg of any group of animals affects not only the manner in which the egg itself differentiates but also the developmental processes that follow fertilization. As we are going to devote our attention to the embryology of the chick which develops from an exceedingly large-yolked egg, we must see how this yolk is accumulated during oögenesis and how its presence modifies the structure of the ovum, so that we may better understand the profound influence which a large yolk-mass exerts on the early growth processes of the embryo.

The part of the hen's egg commonly known as the "yolk" is a single cell, the female sex cell or ovum. Its enormous size as compared with other cells is due to the food material it contains. This food material, or *deutoplasm*, destined to be used by the embryo in its growth, is gradually accumulated within the cytoplasm of the ovum before it is liberated from the ovary. Under the microscope it has the appearance of a viscid fluid in which are suspended granules and globules of various sizes. On chemical analysis yolk is found to contain proteins in various combinations, carbohydrates, and fatty substances in the form of lecithin and sterols as well as neutral fat. Vitamins A, B_1, B_2, D, and E are also present. As these stored materials increase in amount, the nucleus and the cytoplasm are forced toward the surface, so that eventually the deutoplasm comes to occupy nearly the entire cell.

The significance of the chick's liberal endowment of yolk can best be appreciated if one compares the course of its development with that of the frog. The yolk in a frog's egg, though considerable in amount, is not sufficient to carry the embryo through all the changes it must undergo before it attains a body organization like that of the parents. By the time all its yolk has been used the frog embryo has grown only to a larval form, commonly called a tadpole, which is still far short of adult structure. This tadpole is able to secure by its own activities sufficient food to complete its growth and to carry it through the remaining developmental steps which bring it out as a small frog. The chick has no such interrupted embryology. Its generous supply of yolk is sufficient to provide for a rapid and continuous development so that when it emerges from the shell it is already a miniature of its parents.

Figure 8 shows a section of the hen's ovary which includes several young ova and one ovum which is nearly ready for liberation. The very young ova lie deeply embedded in the substance of the ovary. As they accumulate more and more deutoplasm, they crowd toward the surface and finally project from it, maintaining their connection only by a constricted stalk of ovarian tissue. The protuberance containing the ovum is known as an *ovarian follicle*. The bulk of the ovum itself is made up of the yolk. Except in the neighborhood of the nucleus the active cytoplasm is but a thin film enveloping the yolk. About the nucleus a considerable mass of cytoplasm is aggregated. The region of the ovum containing the nucleus and the bulk of the active cytoplasm is known as the *animal pole* because this subsequently becomes the site of greatest protoplasmic activity. The region opposite the animal pole is called the *vegetative*, or *vegetal*, *pole* because, whereas material for growth is drawn from this region, it remains itself relatively less active.

As is the case with the ova of many classes of vertebrates, the ovum of birds has a modified cell membrane called the *zona radiata*. This term was applied to it because, when viewed under high power with a light microscope, it seems to show delicate radial striations. Studied under the greater magnifications possible with an electron microscope this striated appearance is seen to be due to closely packed microvilli. The functional significance of such a structural arrangement is the great increase in membrane surface it affords, thereby enhancing the rate of the metabolic interchanges that can take place at the cell surface.

In a young ovum during the time it is rapidly accumulating deutoplasm the zona radiata is relatively conspicuous. As the amount of accumulated yolk increases, the striated appearance of the cell membrane becomes less

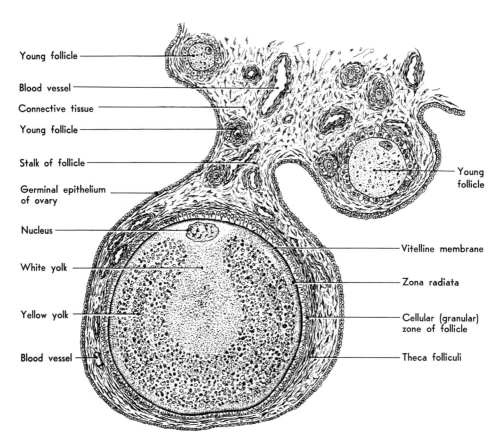

Young follicle

Blood vessel

Connective tissue

Young follicle

Stalk of follicle

Germinal epithelium of ovary

Nucleus

White yolk

Yellow yolk

Blood vessel

Young follicle

Vitelline membrane

Zona radiata

Cellular (granular) zone of follicle

Theca folliculi

Fig. 8.   Diagram showing the structure of a bird ovum still in the ovary. The section shows a follicle containing a nearly mature ovum, together with a small area of the adjacent ovarian tissue. (*Modified from Lillie, after Patterson.*)

evident, and it becomes definitely more robust. This secondarily thickened cell membrane of large-yolked ova is called the *vitelline membrane* (Fig. 8).

Outside the vitelline membrane is an investment of small polygonal cells. These constitute the *cellular* or *granular zone* of the ovarian follicle. This cellular zone is in turn enclosed in a highly vascular layer of connective tissue known as the *theca folliculi* (Fig. 8).

It should be borne in mind that the term "ovum," used without qualification, does not carry a precise significance. It refers always to the female sex cell, but it must be specified whether one means a young ovum befores all the deutoplasm is accumulated, an ovum with its full complement of deutoplasm but unmatured, or a matured ovum ready for fertilization. The more exact terms which parallel those employed in designating the phases of spermatogenesis avoid this difficulty. The ovum during its period of growth and the accumulation of deutoplasm which we have just been considering is a *primary oöcyte*, developmentally homologous to the primary spermato-cyte of the male (Fig. 5). The cells of the granular zone of the ovarian follicle form a protective covering about the oögonium "chosen" for growth into a primary oöcyte and aid in transferring to it the nutriment supplied by the mother from the products of her digested food. This nutritive material is brought in through the blood vessels of the theca, absorbed by the follicular cells, and transferred by them to the growing primary oöcyte where it is elaborated into deutoplasm.

When the full allotment of deutoplasm has accumulated, the investing tissue of the theca is ruptured and the oöcyte is liberated from the ovary. Almost coincidently with *ovulation*, as the discharge of the ovum is called, the first maturation division occurs. In this the nuclear material of the pri-mary oöcyte is halved between the two resulting cells but one of them gets practically all the cytoplasm and its contained food materials (Fig. 9). Both of these unequal cells are *secondary oöcytes*, but only the one which received the entire dowry of deutoplasm has any chance of becoming functional. The small oöcyte containing practically no cytoplasm, because it was budded off from the "animal pole" of the ovum, is commonly called a *"polar body"* or *polocyte*. This polocyte drags along a discouraged existence for a time and may undergo an abortive second maturation division. But it is doomed to degenerate. Again in the second maturation division all the stored food material goes to one cell. The cell which receives no deutoplasm is called the *second polocyte* and is, like the first polocyte, destined for degeneration. The cell which has all the deutoplasm that might have been divided among four sister *oötids* is the mature gamete ready for fertilization (Fig. 9).

If one reviews mentally the phenomena of oögenesis, one is impressed both by the underlying parallelism of oögenesis and spermatogenesis and by certain striking differences between the two processes. The early history of the germ plasm is the same in the two sexes, so too is the period of gamete-mother-cell multiplication, followed by the very definite sequence of a growth phase, which is in turn followed by two maturation divisions (Fig. 5). But each sex shows certain highly characteristic modifications of the pro-cess. Whereas in the male any spermatogonium may produce four mature gametes, in the female three potential oögonia are sacrificed in the more elab-orate provision for a chosen one. For, as a result of the two maturation divi-sions, the oöcyte gives rise to three nonfunctional polar bodies and but one mature gamete which gets all the food accumulated by the primary oöcyte in its long growth period. The very liberality of the stored food material col-

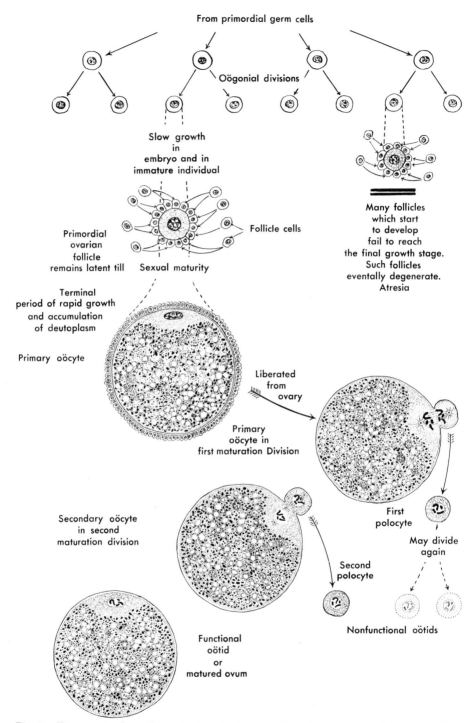

From primordial germ cells

Oögonial divisions

Slow growth
in
embryo and in
immature individual

Many follicles
which start
to develop
fail to reach
the final growth stage.
Such follicles
eventually degenerate.
Atresia

Primordial
ovarian
follicle
remains latent till

Follicle cells

Sexual maturity

Terminal
period of rapid growth
and accumulation
of deutoplasm

Primary oöcyte

Liberated
from
ovary

Primary
oöcyte in
first maturation Division

First
polocyte

May divide
again

Secondary oöcyte
in second
maturation division

Second
polocyte

Nonfunctional oötids

Functional
oötid
or
matured ovum

Fig. 9. Diagram to show the main steps in the growth and maturation of the ovum. For the sake of simplicity the species number of chromosomes is assumed to be eight.

lected in the maturing ovum renders it inactive and passively dependent on surrounding structures for its transportation. In contrast, the concentration of nuclear material in their small, compact heads and the development of an elaborate propulsive mechanism give to the spermatozoa a high degree of independent motility.

Significance of Maturation   The events in the maturation of the male and female gametes which have just been discussed are but the more evident phases of the process. There have been changes of profound significance going on at the same time in the nuclear material. It would carry us far afield into cytology and genetics to discuss these changes in detail and attempt an interpretation of their full meaning. But we can indicate briefly wherein their importance lies.

It has already been stated that the inheritance of an individual comes to it by way of the gametes arising from the germ plasm of its parents. We can be more definite. It comes by way of the chromosomes in the nuclei of the gametes. The chromosomal complement of the nuclei in the cells which go to make up the body of an individual is definite and constant. The elaborate mechanism of mitosis divides the chromosomes lengthwise into qualitatively and quantitatively equivalent daughter chromosomes. Thus, in each of the countless cell divisions involved in the growth of an individual, the number of chromosomes remains in the daughter cells what it was in the parent cell. The chromosome number so maintained is different in different species, but in the various cells of individuals of the same species it is fixed and definite, the "species number of chromosomes" in the parlance of cytologists and geneticists.

Moreover in each cell division, when the chromosomes take shape from the scattered chromatin granules of the resting nucleus, they reappear in the same characteristic shapes and sizes exhibited by the chromosomes of the previous cell division. This individuality of the chromosomes is very likely to be overlooked by the uninitiated, because often schematic diagrams illustrating mitosis, for the sake of simplicity, show the chromosomes as if they were all exactly alike. But a cytologist, when he has studied the cell structure of an animal intensively, will be able to tell us not only how many chromosomes we can count on finding at each mitosis but also how each chromosome will look. One he will characterize as long and bent, another as slender and straight, still another as short and plump, and so on. Furthermore on careful study of these chromosomes one finds that they are present in pairs, the members of which are similar in size and shape. The members of a pair are not usually located next to one another on the spindle of an ordinary somatic mitosis, but methodical comparison of their size and shape enables the cytologist to chart the chromosomes of a cell, similar pair, by similar pair (Fig. 10).

The fact that the chromosomes thus occur in recognizable couples is of great significance to the geneticist in his study of the manner in which hereditary traits are passed on from one generation to the next. We are already familiar with the fact that all the cells of an animal's body have been formed by the division and redivision of the fertilized egg cell or zygote. We know that in each of these mitotic divisions each chromosome is split in half so that the daughter cells receive the same number of chromosomes that the parent cell contained. Obviously then, if a cell taken from a rooster's comb or anywhere else in his body showed 36 chromosomes, the zygote from which the rooster grew inevitably had 36 chromosomes. Now if we study the formation of the zygote, we find that of its full species number of chromosomes half were brought in by the male gamete and half by the female gamete. And if we study the individual shapes and sizes of their chromosomes we find that each gamete contributed one member of each pair of chromosomes which characteristically appear in the cells of the body in that species.

Much work has been done by geneticists and cytologists in an effort to link specific bodily traits with specific chromosomal structure. It is now quite generally believed that the material responsible for transmitting hereditary traits is strung out along the length of the chromosomes in more or less

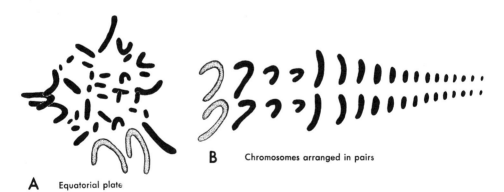

B   Chromosomes arranged in pairs

A   Equatorial plate

Fig. 10.   The chromosomes of the fowl. The sex chromosomes are indicated by stippling. (A) Actual appearance (camera lucida drawing, polar view) of the chromosomes as they lie in the equatorial plane of a mitotic spindle just before the beginning of the metaphase splitting. (B) Diagram, to the same scale as A, showing how the chromosomes may be arranged in pairs, the members of which are similar to each other in size and shape. [After Hance, J. Morphol. Physiol., **43** (1926), slightly modified as to arrangement.]

segregated units. These units, like the atoms of a chemical substance, are too small to be identified even with the aid of the microscope, but under experimental procedures they can, nevertheless, be made to give evidence of their presence. So, even though they cannot be seen, it is convenient to have a concise way of designating them, and they are spoken of as "genes," with perhaps a more casual air of familiarity than our real acquaintance with them justifies.

One way in which evidence of gene arrangement within chromosomes may be secured has been extensively used for many years. When in an experimental colony of animals of known stock there appears an individual which differs in some definite physical respect from its parents, it is called a *mutant*. If physical peculiarities depend on genes within chromosomes, theoretically it should be possible by critical study of the dividing germ cells of the mutant to find atypical chromosomes. There should be some change from the usual shape due to a gene not usually present, or a minute gap where a gene usually present is missing. In some of the experimental forms, such as the fruit fly, Drosophila, which has been intensively studied from the genetic standpoint, this has proved to be the case. Even more striking than the evidence obtained from spontaneous mutants has been that obtained more recently by irradiation of the germ cells. If during their maturation germ cells are subjected to x-rays of an intensity not quite sufficient to kill them, some of the genes are destroyed or displaced. When animals so treated are bred, some of their offspring show definite changes in bodily structure. In other words mutations have been artificially induced which may be correlated with equally definite changes in the gene composition of particular chromosomes. The work of thus tying up specific hereditary traits with definite chromosomes, though as yet in its infancy, has already yielded most interesting results. We are beginning to see more clearly the manner in which both the racial heritage and the particular traits of an individual are passed on. The links in the endless chain of heredity are the gene-bearing chromosomes — a definite number of pairs maintained constant in all the cells of an individual by mitosis; the same definite number of pairs in the cells of each succeeding generation reestablished when each gamete brings in one member of the new pair.

The final two divisions in spermatogenesis and oögenesis are not typical mitotic divisions. To distinguish them they are called *meiotic divisions*, or less technically, *maturation divisions*. In these divisions the chromosomes of the gametes are reduced to half the species number. The process of thus halving the species number of chromosomes is called *meiosis*. Cytologists have worked out the mechanism of these maturation divisions with great care in many forms. In some forms chromosome reduction takes place in the first, in others in the second maturation division. Which of the two divisions is a reduction division in any particular case is of minor significance. The reduction itself is the thing of interest. Stripped of detail and without reference to many peculiar modifications encountered in different animals, a *reduction division* is a cell division in which the chromosomes are not split in the metaphase. Instead the chromosomes are redistributed, half of them moving bodily to one daughter cell and half to the other.

In this halving process the way is paved for separating the chromosomal pairs by a special mechanism termed *synapsis* which occurs in the prophase of a maturation division but not in an ordinary mitosis. If we look at the spindle of an ordinary mitosis (Fig. 11, *B*) we can recognize the members of the chromosomal pairs by their size and shape, but the members of the pairs do not lie next to one another. When, during the prophase, the spireme thread was broken up into chromosomes these chromosomes aggregated at once in haphazard arrangement at the equator of the spindle. In the prolonged prophase of the first maturation division the members of the chromosomal pairs come to lie close to each other and so remain for some time (Fig. 11, *E*). This pairing off of the chromosomes in the prophase is what is meant by the term synapsis. These synaptic pairs of chromosomes, still in intimate association, finally move to the equator of the spindle (Fig. 11, *F*). Whereas in an ordinary mitosis splitting of the chromosomes would occur when they had arrived at the equatorial position, in a reduction division the two members of the synaptic pair are separated from each other, one going bodily to either pole of the spindle (Fig. 11, *G*, *H*). The resulting cells thus receive half the species number of chromosomes, and this half-complement (*haploid number*) is made up of one member of each of the pairs characteristically present in the species (*diploid number*). If the first of the maturation divisions has been a reduction division, as just described, then the second maturation division will be an *equation division*. In it the chromosomes split longitudinally yielding, as the designation equation division implies, two daughter cells, each with the same haploid number of chromosomes exhibited at the beginning of the division (Fig. 11, *I–L*).

In many forms which have been carefully studied the synaptic pairs of chromosomes, as the first maturation division is approached, exhibit a peculiar quadripartite appearance (Fig. 11, *E*, *F*). This has led to their being designated as a *tetrad*. This condition is due merely to the fact that internal splitting of the chromosomes has become apparent before their parts are widely separated by movement toward the poles of the spindle. Thus a tetrad is nothing but a synaptic pair of chromosomes in which each member of the

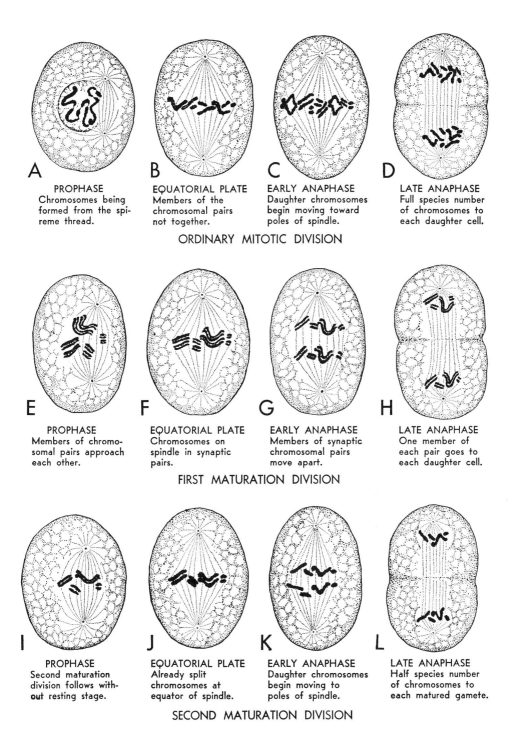

A
PROPHASE
Chromosomes being
formed from the spi-
reme thread.

B
EQUATORIAL PLATE
Members of the
chromosomal pairs
not together.

C
EARLY ANAPHASE
Daughter chromosomes
begin moving toward
poles of spindle.

D
LATE ANAPHASE
Full species number
of chromosomes to
each daughter cell.

ORDINARY MITOTIC DIVISION

E
PROPHASE
Members of chromo-
somal pairs approach
each other.

F
EQUATORIAL PLATE
Chromosomes on
spindle in synaptic
pairs.

G
EARLY ANAPHASE
Members of synaptic
chromosomal pairs
move apart.

H
LATE ANAPHASE
One member of
each pair goes to
each daughter cell.

FIRST MATURATION DIVISION

I
PROPHASE
Second maturation
division follows with-
out resting stage.

J
EQUATORIAL PLATE
Already split
chromosomes at
equator of spindle.

K
EARLY ANAPHASE
Daughter chromosomes
begin moving to
poles of spindle.

L
LATE ANAPHASE
Half species number
of chromosomes to
each matured gamete.

SECOND MATURATION DIVISION

Fig. 11. Diagrams showing schematically the difference between a reduction division in the process of maturation and an ordinary mitotic division. For the sake of simplicity of representation the species number of chromosomes is assumed to be eight.

pair is showing a sort of precocious internal separation. This same type of early division can be observed in ordinary mitoses but it is not usually as conspicuous as in the first maturation division.

If we recall, now, what we learned of the gene composition of chromosomes, it will be evident that the cells formed in the maturation divisions contain different hereditary potentialities because they contain different chromosomes, not halves of the same chromosomes as results in an ordinary mitosis. What hereditary possibilities are discarded into the polar bodies thrown off from the femal gamete and what are retained in the mature ovum is a matter of chance distribution. What potentialities find their way into the particular sperm which alone, out of millions of its fellows, fertilizes the ovum is likewise fortuitous.

"Thus in the game of life, the maturation processes virtually shuffle the hereditary pack and deal out half a 'hand' to each gamete. A full hand is obtained by drawing a partner from the 'board,'—by combining with some other gamete of the opposite sex. Hence offspring resemble their parents because they play the game of life with the same kind of cards, but not, however, with the same hands. The minor differences in offspring, or the variations from the standard type that always go with these basic resemblances, are due to variations in the distribution of the genes during maturation" and new combinations made in fertilization.

Thus there is produced sufficient stability to ensure continuity and at the same time sufficient variety to ensure progress. "For the offspring will in the main resemble progenitors which have successfully lived in the prevailing conditions of the past, but will exhibit sufficient variability among themselves to insure that some of them shall successfully live in any conditions likely to arise in the future."[1]

Sex Determination   Reference has already been made to the fact that cytologists and geneticists have begun to correlate peculiarities of individual chromosomes with bodily characteristics. It now seems quite clear that sex, also, is influenced by the particular chromosomal combination which occurs with the fusion of the two gametes. Just before the turn of the present century Henking, in studying the chromosomal pattern of certain insects, was impressed with the fact that there was one pair of chromosomes which lagged behind the others in moving toward the poles of the spindle in the maturation divisions. This was intriguing, but its significance was not at first ap-

[1] From William Patten: Life, Evolution, and Heredity, *Sci. Monthly*, **21:** 122–134.

parent. After the manner of mathematicians in giving the "unknowns" in their problems the last letters in the alphabet as noncommittal designations, the members of this chromosomal pair were christened X and Y. It was several years later that McClung and Wilson came to the conclusion that this pair of chromosomes was involved in sex determination. In line with this pioneering work, practically all the animals which have been critically studied show a slight but strikingly consistent difference in the chromosome picture exhibited by the cells in the bodies and in the germ line of the two sexes. In one sex, the cells have all their pairs of chromosomes symmetrically mated. In the opposite sex one of the pairs of chromosomes is likely to be peculiar in that its members are quite different in size and shape instead of being like each other as is the general rule. Subsequent experimental work has supported the conclusions of McClung and Wilson and given us considerable evidence indicating that the reason the members of the X-Y pair are unlike is because something that makes for femaleness is carried in the X chromosome and absent in the Y.

We have seen that in the reduction division of the gametes the members of the chromosomal pairs are separated (Fig. 11). If the X chromosome does carry whatever makes for femaleness and its Y mate lacks this or carries an antagonistic male determining factor, it is evident that when in a reduction division the members of this chromosomal couple go to opposite poles of the spindle the resulting cells will carry opposite sex tendencies (Fig. 12).

Assuming (as actually is the case in most of the forms so far studied[1]) that it is the female which has all its chromosomes symmetrically paired and the male that has an unbalanced pair, the chromosomal combinations taking place in fertilization can readily be depicted graphically (Fig. 13). When the female gametes mature, all of them will appear alike as far as chromosomes are concerned, because in maturation one member of each balanced pair goes to each cell (Fig. 11, *E–L*). The reduction division in spermatogenesis, however, separates the unbalanced X-Y pair and consequently half the gametes receive an X chromosome and the other half a Y chromosome (Fig. 12). The chances of an X-carrying gamete uniting with an ovum and giving rise to a female are the same as those of a Y-bearing chromosome fertilizing the ovum to produce a male (Fig. 13). Thus this theory accounts at the same time for the approximate equality of the numbers of males and females produced,

---

[1] Recent investigations indicate that the fowl probably is one of the exceptions to the general rule, that it is the male which produces two types of gametes or, to use the technical expression, exhibits digamety. Apparently the hen produces ova which differ in sex-determining potentialities, while the cock produces spermia which are all alike in this respect. The interpretation given in the text can readily be inverted to cover female digamety although it is outlined concretely for the more usual condition of male digamety.

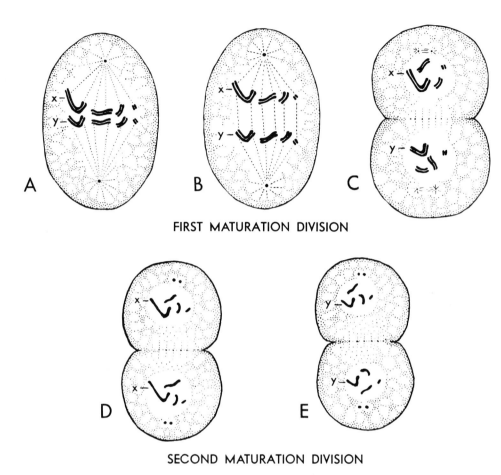

FIRST MATURATION DIVISION

SECOND MATURATION DIVISION

Fig. 12. Schematic diagrams indicating the manner in which the chromosomes believed to be involved in sex determination are distributed in the maturation divisions.

and for the observed differences in the chromosome picture presented by the two sexes.

Recent experimental evidence seems to indicate that there are other factors involved in sex determination, and that sex may not be as irrevocably fixed at the time of fertilization as was at first believed. There seems little doubt that the chromosomal combination is effective in starting the trend of development toward one sex or the other, but apparently this trend may be inhibited or modified by massive doses of male or female sex hormones during the period when the gonads are being differentiated. Moreover, from his work on intersexes, Bridges believes that there are genes in other chromosomes than the X-Y pair which influence the development of maleness or

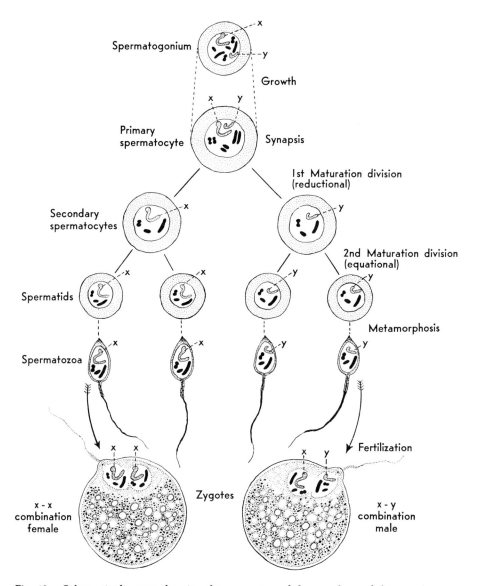

Fig. 13.   Schematic diagram showing the separation of the members of the sex chromosome pair in maturation and their recombinations in fertilization. It is assumed that the
species number of chromosomes is eight and that it is the male which produces gametes
of different potentialities with regard to sex determination. The sex chromosomes are
stippled; other chromosomes are drawn in solid black.

femaleness. When these factors are in balance, the XY or XX combination is sufficient to throw the course of development to one sex or the other. Unbalance of the ancillary factors from other chromosomes under certain circumstances may operate against this determining influence of the so-called "sex chromosomes" with the resultant production of sex intergrades. Whatever future work may show the accessory or modifying factors to be, it seems quite clearly established that the X-Y chromosomal pair plays an important, probably the major, role in the determination of sex.

In 1949 Barr and Bertram first presented evidence that it was possible to see a sexual difference in the fixed and stained nuclei of nondividing somatic cells. They found that in the nuclei of cells from females there was usually a conspicuous, characteristically located mass of chromatin that was absent in the nuclei of cells taken from a male (Fig. 14). This chromatin mass is called the *sex chromatin*. Caution must be used in arriving at a conclusion that sex chromatin is absent on the basis of the study of only a few nuclei. Good histological preparations are so thin that they frequently include only part of a nucleus, and the sex chromatin may well be in the part of the nucleus not included in a particular section. Nevertheless, when based on the critical study of any considerable number of nuclei, consistent absence of sex chromatin is reliably indicative of male chromosomal status. The fact that the presence or absence of sex chromatin can be demonstrated in cells readily obtainable from simple biopsies or even from cells scraped from the oral epithelium makes the finding of particular value. Previously a cytological diagnosis of the genetic sex of cells rested on the time-consuming process of culturing the cells and determining their chromosomal constitution by

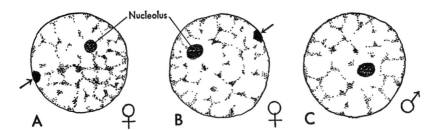

Fig. 14.   Drawings of the nuclei of human epidermal cells to show the sex chromatin. (*A*) and (*B*) Nuclei from a female, the sex chromatin indicated by arrows. (*C*) Nucleus from a male, showing the absence of sex chromatin. Note that the nucleolus, although eccentric in position, does not lie against the nuclear membrane. [*Schematized from the work of Moore, Grahm, and Barr, Surg., Gyn. and Obs.*, 96 (1953).]

making counts on those caught in the metaphase of mitosis. The full story of the origin of the sex chromatin is not yet known, but it is believed to be derived in some manner from the X chromosomes.

Fertilization   Direct observations of fertilization in such forms as birds and mammals where the process occurs inside the body of the female are exceedingly fragmentary. Fortunately much more information is available from the study of water-living animals in which the eggs are commonly deposited in the water outside the maternal body and there fertilized. When such eggs are placed in a watch glass and sperm introduced, the whole process of fertilization can be studied under the microscope. Interpreting in the light of such observations we can piece together a generalized account of the process of fertilization which in all probability is applicable without essential modification to birds or mammals.

When liberated from the ovary, an ovum at once begins to undergo certain changes which can be characterized as aging or deterioration. Among other things there is a tendency for its protoplasm to become progressively more coarse. With this loss of the originally finely dispersed phase of its colloidal material the ovum loses vigor. These changes progress rapidly to a point where the ovum, although technically still alive, can no longer be fertilized. If, however, the ovum is fertilized reasonably promptly after its liberation, these deteriorative changes are checked and the protoplasm increases its activity in a way that is often described as "being rejuvenated." The nature of these changes is not as yet fully understood, but they involve increase in permeability and increase in oxidation rate. There is also an increase in the amount of ammonia excreted. This indicates that there has been an increase in purine metabolism, which is important in the synthesis of nuclear materials. These changes symbolize, of course, the beginning of the period of tremendously rapid growth which is destined to end in the production of a new individual.

Interestingly enough, fertilization in the usual manner by the male sex cell is not the only way an ovum may have its deterioration checked and be started on its growth phase. A variety of other stimuli may be substituted for the male sex cell. Changes in the ionic concentration of the sea water is effective in initiating development in the eggs of some marine invertebrates. Insect eggs have been started developing by stroking them with a camel's hair brush. Pricking with a needle has been effective in inducing frog's eggs to begin to grow. The experimental initiation of development in an egg by means other than its union with the male sex cell is known as *artificial par-*

thenogenesis. The fact that artificial parthenogenesis is possible emphasizes the fact that all the necessary morphogenetic factors for development are present in the ovum and that the process of fertilization is essentially, as Barth puts it, a sort of "release mechanism." In other words something from the sperm cell, or in artificial parthenogenesis the chemical or mechanical factor employed, releases potential energy already present in the egg. It should not be overlooked, however, that this release mechanism is far from the whole story. The sperm brings with it its quota of hereditary potentialities, and the mixture of germinal material from two parents produces a vigor of growth not manifested in eggs starting their development by artificial parthenogenesis.

Turning now from the ovum to the male sex cells, it is interesting that in the testes and sperm ducts the spermatozoa show but little indication of their power of locomotion. During coitus, however, certain accessory glands opening into the sperm ducts discharge a considerable amount of fluid in which the spermia are floated and in which they become actively motile. When this fluid (*semen*) has been deposited in the genital tract of the female, the spermia are under conditions not radically unlike the spermia of a fish discharged into sea water near eggs recently laid by the female. The bird ovum, recently liberated from the ovary, is awaiting the spermia in the upper part of the oviduct. Here the ovum becomes surrounded by swarms of spermatozoa which have made their way thither by their own active locomotion through the lower part of the genital tract.

The process to this point is known as *insemination*. Still other preliminary steps are to be passed through before the nuclei of the male and female gametes unite to complete the process of fertilization. The spermia once having reached the neighborhood of the ovum tend to remain there held by chemical interactions which are not as yet fully understood.

Apparently, at least in the case of some of the most critically studied of the marine invertebrates, both the ova and the sperm produce substances having definite effects on each other. These substances are in some ways similar to hormones, and because they are produced by gametes, they have been called *gamones*. The active substances produced by ova are known as *gynogamones* and the comparable substances resident in sperm are called *androgamones*. Work in this field is relatively new, and the conclusions from it must still be regarded as tentative. On the basis of current interpretations if appears as if there were two gynogamones. Gynoganome I is believed to activate the sperm cells to vigorous swimming movements. Gynogamone II makes the surface of the spermatozoa heads sticky, and may thus be a factor in adherence of a sperm cell to the surface of the egg cell for sufficient time

to initiate penetration. The fact that it also causes sperm cells to stick together, or agglutinate, seems incidental.

There seem to be, also, two androgamones. Androgamone I is believed
to inhibit active sperm movement. Spermatozoa contain a minimal amount
of energy-producing materials so that their potential kinetic energy is strictly
limited. Inactivity until the critical time arrives is, therefore, important and
it appears as if androgamone I enforces such inactivity until its action is
nullified by the activating effects, first of the male accessory gland secretions
and then by gynogamone I. Androgamone II appears to be responsible for
the dissolving of the outer egg membranes, thereby facilitating the penetration of sperm cells.

When a sperm cell has penetrated the outer investment of the egg, a
minute projection of ovarian cytoplasm, called the fertilization cone, rises
up to meet it and draws it into the ovum (Fig. 15, *A*, *B*). Not infrequently,
especially in birds, several spermatozoa penetrate the ovum, but since only

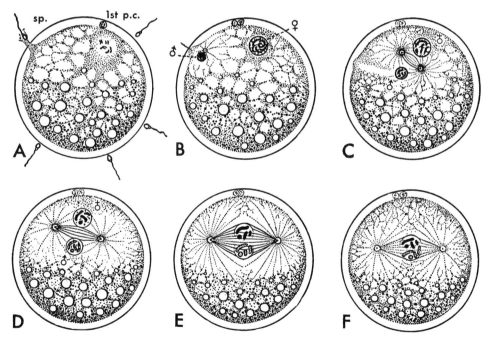

Fig. 15.   Diagrams schematically illustrating the process of fertilization and the formation of
the first cleavage spindle. As in Fig. 11, the species number of chromosomes is assumed to be
eight.

Abbreviations: sp., spermium; *p.c.*, polar cell (polar body). The male and female symbols designate the male and female pronuclei, respectively. (*After William Patten.*)

one of these takes part in fertilization we can neglect the others. Once a sperm has penetrated the ovum the peripheral cytoplasm produces a clear viscid substance which adheres to the inner face of the vitelline membrane, thickening it into the so-called *fertilization membrane*. The fact that no spermia penetrate the ovum after the fertilization membrane has been formed may be due as much to chemical changes in the ovum following the entrance of the sperm as to the mechanical impediment offered by the membrane.

The first maturation division of the ovum has usually occurred at about the time of ovulation. The second maturation division is very likely not to occur until after the ovum has been penetrated by a spermium. While this final maturation division is being completed by the ovum, the spermium undergoes very striking changes. Its tail is usually dropped off when it enters the ovum. Once within the ovarian cytoplasm the sperm head increases rapidly in size and its chromosomal contents again become distinctly recognizable (Fig. 15, *B–D*). In this condition it is called the *male pronucleus*, and the nucleus of the ovum after the second maturation division has been completed is called the *female pronucleus*. Meanwhile the centrosomal apparatus, which was carried in the spermium where its tail was attached to its head, becomes much more conspicuous. As it approaches the female pronucleus, the male pronucleus with its associated centrosome rotates so that the centrosome moves ahead of the chromatin material (Fig. 15, *B*). By the time the two pronuclei are close to each other this centrosome has divided and formed a mitotic spindle on which the chromosomes brought in by the sperm and those in the ovum both aggregate (Fig. 15, *E, F*). Fertilization can be regarded as complete when the maternal and paternal chromosomes are thus grouped together ready to be split in the impending first cell division in the life of the new individual.

Formation of the Accessory Coverings of the Ovum   Fertilization, as we have seen, normally takes place just as the ovum is entering the oviduct. The accessory coverings, as the albumen, shell membrane, and shell are called, are secreted about the ovum during its subsequent passage toward the cloaca. In the part of the oviduct adjacent to the ovary a mass of stringy *albumen* is produced. This adheres closely to the vitelline membrane and projects beyond it in two masses extending in either direction along the oviduct. Due to the spirally arranged folds in the walls of the oviduct, the egg as it moves toward the cloaca is rotated. This rotation twists the adherent albumen into the form of spiral strands projecting at either end of the yolk, known as the *chalazae* (Fig. 16). Additional albumen, which has been se-

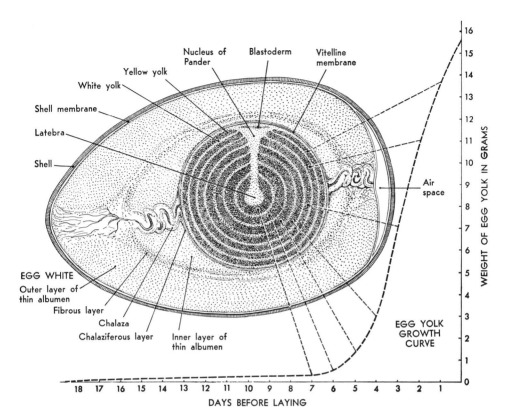

Fig. 16. Diagram showing the structure of the hen's egg at the time of laying. The graph indicates the rate of growth of the egg during the 18 days preceding its laying. The lines leading from the various layers of the yolk to the growth curve emphasize the time at which these layers were formed. (*Redrawn, with slight modifications, after Witschi, "Development of Vertebrates," W. B. Saunders Company, Philadelphia, 1956.*)

creted abundantly in advance of the ovum by the glandular lining of the oviduct, is caught in the chalazae and during the further descent of the ovum is wrapped about it in concentric layers. These lamellae of albumen may be easily demonstrated in an egg which has had the albumen coagulated by boiling. The albumen-secreting region of the oviduct constitutes about one-half of its entire length.

The *shell membranes* which consist of sheets of matted organic fibers are added farther along in the oviduct. The *shell* is secreted as the egg is passing through the shell-gland portion of the oviduct. The entire passage of the ovum from the time of its discharge from the ovary to the time when it is ready for laying has been estimated to occupy about 22 hours. If the completely formed egg reaches the cloacal end of the oviduct during the middle of the day it is usually laid at once, otherwise it is likely to be retained until the following day. This overnight retention of the egg is one of the factors which accounts for the variability in the stage of development reached at the time of laying.

Structure of the Egg at the Time of Laying   The arrangement of structures in the egg at the time of laying is shown in Fig. 16. Most of the gross relationships are already familiar because they appear so clearly in eggs which have been boiled. If a newly laid egg is allowed to float free in water until it comes to rest and is then opened by cutting away the part of the shell which lies uppermost, a circular whitish area will be seen to lie atop the yolk. In eggs which have been fertilized, this area is somewhat different in appearance and noticeably larger than it is in unfertilized eggs. The differences are due to the development which has taken place in fertilized eggs during their passage through the oviduct. The aggregation of cells which in fertilized eggs lies in this area is known as the *blastoderm*. The structure of the blastoderm and the manner in which it grows will be taken up in Chap. 4.

Close examination of the yolk will show that it is not uniform throughout either in color or in texture. Two kinds of yolk can be differentiated, *white yolk* and *yellow yolk*. Aside from the difference in color visible to the unaided eye, microscopical examination will show that there are differences in the granules and globules of the two types of yolk, those in the white yolk being in general smaller and less uniform in appearance. The principal accumulation of white yolk lies in a central flask-shaped area, the *latebra*, which extends toward the blastoderm and flares out under it into a mass known as the *nucleus of Pander*. In addition to the latebra and the nucleus of Pander there are thin concentric layers of white yolk between which lie much thicker layers of yellow yolk. The concentric layers of white and yellow yolk are said to indicate the daily accumulation of deutoplasm during the final stages in the formation of the egg. The outermost yolk immediately under the vitelline membrane is always of the white variety.

The albumen, except for the chalazae, is nearly homogeneous in appearance, but near the yolk it is somewhat more dense than it is peripherally. The chalazae serve to suspend the yolk in the albumen.

The two layers of shell membrane lie in contact everywhere except at the large end of the egg where the inner and outer membranes are separated to form an *air space*. This space is stated (Kaupp) to appear only after the egg has been laid and cooled from the body temperature of the hen (about

106° F) to ordinary temperatures. In eggs which have been kept for any length of time, the air space increases in size due to evaporation of part of the water content of the egg. This fact is taken advantage of in the familiar method of testing the freshness of eggs by "floating them."

The egg shell is composed largely of calcareous salts. These salts are derived from the food of the mother, and if lime-containing substances are not furnished in her diet, the shell is defectively formed or even altogether wanting. The shell is porous allowing the embryo to carry on exchange of gases with the outside air by means of specialized vascular membranes arising in connection with the embryo but lying outside it, directly beneath the shell.

Incubation   When an egg has been laid, development ceases unless the temperature of the egg is kept nearly up to the body temperature of the mother. Moderate cooling of the egg does not, however, result in the death of the embryo. It may resume its development if it is brooded by the hen or artificially incubated even after the egg has been kept for several days at ordinary temperatures.

The normal incubation temperature is that at which the egg is maintained by the body heat from the brood-hen. This is somewhat below the blood heat of the hen (106° F). When an egg is allowed to remain undisturbed, the yolk rotates so that the developing embryo lies uppermost. Its position is then such that it gets the full benefit of the warmth of the mother.

In incubating eggs artificially the incubators are usually regulated for a heat of 99 to 100° F (36 to 38° C). At this temperature the chick is ready for hatching on the 21st day. Development will go on at considerably lower temperatures, but its rate is retarded in proportion to the lowering of the temperature. Below about 21° C development ceases altogether.

If eggs which have been cooled after laying are to be incubated for the purpose of securing embryos of a particular stage of development, 3 or 4 hours are ordinarily allowed for the egg to become warmed to the point at which development begins again. For example if an embryo of "24-hours' incubation age" is desired, an egg that has been subjected to cooling should be allowed remain in the incubator about 27 hours. Even with allowance made for the warming of the egg and with exact regulation of the temperature of the incubator, the stage of development attained in a given incubation time will vary widely in different eggs. The factor of individual variability, which must always be reckoned with in developmental processes, undoubtedly accounts for some of the variation. The different lengths of time occupied by eggs in traversing the oviduct, the overnight retention of eggs not ready for laying until toward sundown, and especially the varying time different eggs have been brooded before being removed from the nest account for further variations. The designation of the age of chicks in hours of incubation is, therefore, not exact, but merely a convenient approximation of the average condition reached in that incubation time.

CHAPTER 3

THE PROCESS OF CLEAVAGE

Effect of Yolk on Cleavage Immediately after its fertilization the ovum enters upon a series of mitotic divisions which occur in close succession. This series of divisions constitutes the process of *cleavage* or *segmentation*. In birds cleavage takes place before the egg is laid, during the time it is traversing the oviduct.

A mitotic division, whether it be a cleavage division of the ovum or the division of some other cell, is carried out by the active protoplasm of the cell. The food material stored in an egg cell as deutoplasm is nonliving and inert. The deutoplasm plays no part in mitosis except as its mass mechanically influences the activities of the protoplasm of the cell. It is obvious that any considerable amount of yolk will retard the division, or prevent the complete division, of the fertilized ovum. The amount and distribution of the yolk will therefore determine the type of cleavage.

An egg such as that of Amphioxus which has a scanty amount of yolk, fairly uniformly distributed throughout the cytoplasm, is called *isolecithal* (*homolecithal*). An isolecithal egg undergoes a type of cleavage which is essentially an unmodified mitosis. The yolk is not sufficient in amount, nor sufficiently localized, to alter the usual mode of cell division. These early divisions follow each other in very rapid sequence, so that cells have little time to grow between mitotic divisions. This means that a rapidly growing young embryo, paradoxically, has smaller and smaller cells (Fig. 17).

In Amphibia the ovum contains a considerable amount of yolk, and the accumulation of the yolk at one pole has crowded the nucleus and active cytoplasm of the ovum toward the opposite pole. An egg in which the yolk is thus concentrated at one pole is termed *telolecithal*. Cleavage in such an egg is initiated at the animal pole where the nucleus is located together with most of the active cytoplasm. The division of the nucleus is a typical mitotic division. The division of the cytoplasm is effected rapidly at the animal pole of the egg where the active cytoplasm is aggregated. When, however, the yolk-mass is encountered, the process is greatly retarded. So slowly, in fact, is the division of the yolk accomplished, that succeeding cell divisions begin at the animal pole of the egg before the first cleavage is completed at the vegetative pole. When the yolk thus impedes division at the vegetative pole, the

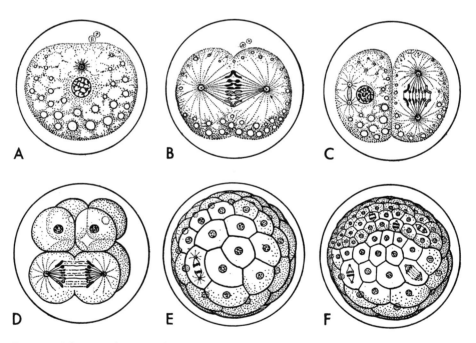

Fig. 17. Schematic diagrams showing the general sequence of events in the cleavage of an egg having but a small amount of yolk in its cytoplasm. (*After William Patten.*)

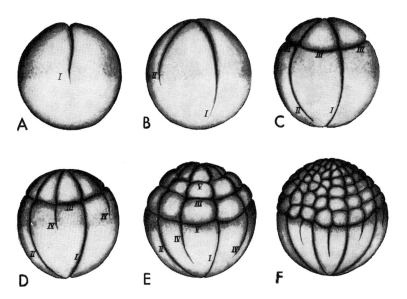

Fig. 18. Cleavage in the frog's egg. The darkness of the upper hemisphere is due to the presence of pigment in the cytoplasmic cap, while the lighter appearance of the lower hemisphere is caused by the massing of yolk in that part of the egg. The cleavage furrows are designated by Roman numerals indicating the order of their appearance. In *A* and *B* note the retarding effect of the yolk on the extension of the cleavage furrows toward the vegetative pole. In *C* observe the displacement of the third cleavage furrow away from the yolk-laden pole, toward the center of the mass of active cytoplasm. The displacement of the center of activity from the geometrical center of the egg and the mechanical retardation of cleavage at the vegetative pole—both due to the yolk-mass—result in the formation of a morula with many small blastomeres in the animal (apical) hemisphere and fewer large blastomeres in the vegetative (abapical)hemisphere.

cells of this hemisphere remain much larger than those at the animal pole because in a given time they have not divided as often (Fig. 18). Later we shall see that the moulding effect of the yolk on embryonic configuration is by no means limited to the period of cleavage but, on the contrary, that its effects are manifest well into the period of organ formation. Meanwhile the obvious modifying effect on cleavage exerted by the moderate amount of yolk present in the eggs of such forms as the frog indicates the profound effect we can expect the much greater yolk-mass of the hen's egg to have on the course of chick development.

Sequence and Orientation of the Cleavage Divisions in Birds   Cleavage in bird's eggs begins as it does in the eggs of Amphibia (Fig. 18), but the mass of the inert yolk material is so great that the yolk is not divided. The process of segmentation is limited to the small disk of protoplasm lying on the surface of the yolk at the animal pole, and is for this reason referred to as *discoidal cleavage* (Fig. 19). The fact that the whole egg is not divided is indicated by designating the process as partial (*meroblastic*) cleavage in distinction to the complete cleavage (*holoblastic*) seen in eggs containing less yolk. The cells formed in the process of cleavage are known as *blastomeres* whether they are completely separated, as is the case in holoblastic cleavage, or only partially separated, as in meroblastic cleavage.

In the bird's egg which is about to undergo cleavage, the disk of active protoplasm at the animal pole (*blastodisk*) is a whitish, circular area about 3 mm in diameter. The central portion of the blastodisk is surrounded by a somewhat darker appearing marginal area known as the *periblast*. The protoplasm of the blastodisk, especially in the periblast region, blends into the underlying white yolk so that is it difficult to make out any line of demarcation between the two. It is in the central area of the blastodisk that cleavage furrows first appear. Neither the nuclei resulting from the early cleavages nor the cleavage furrows invade the marginal periblast until very late in the process of segmentation.

The nature of the series of divisions in the meroblastic, discoidal cleavage characteristic of the eggs of birds is, as has already been pointed out, predetermined to a great extent by the amount and distribution of the yolk. Another determining factor is the tendency of mitotic spindles to develop so that the long axis of the spindle lies in the axis of the greatest dimension of the unmodified cytoplasm. The cleavage furrow always forms at right angles to the long axis of the mitotic spindle. Figure 19 shows the succession of the cleavage divisions in the egg of the pigeon. The diagrams represent surface views of the blastodisk and an area of the surrounding yolk, the shell and albumen having been removed. The first cleavage furrow [Fig. 19, *A*(I–I)] cuts into the egg in a plane coinciding with the axis passing through the animal pole and the vegetative pole.

In each of the two blastomeres resulting from the first cleavage division, mitotic spindles initiating the second cleavage arise at right angles to the position which was occupied by the first cleavage spindle (Fig. 20, cf. *A* and *B*). This determines that the two simultaneously appearing second cleavage furrows will be at right angles to the first. Since these two second cleavage furrows lie in the same plane and are apparently continuous, they are usually considered together. A very good way of acquiring a clear conception of the

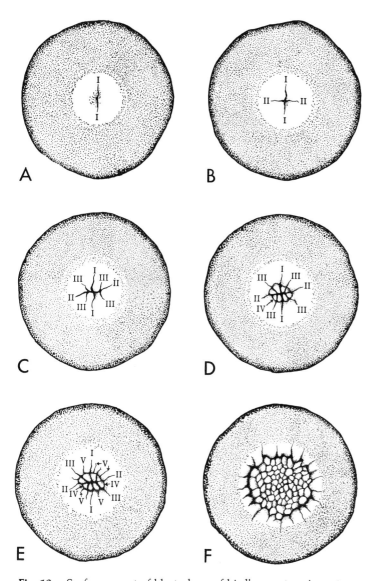

Fig. 19. Surface aspect of blastoderm of bird's egg at various stages of cleavage. The blastoderm and the immediately surrounding yolk are viewed directly from the animal pole, the shell and albumen having been removed. The order in which the cleavage furrows have appeared is indicated on the diagrams by Roman numerals. (A) First cleavage; (B) second cleavage; (C) third cleavage; (D) fourth cleavage; (E) fifth cleavage; (F) early morula. (*Based on Blount's photomicrographs of the pigeon's egg.*)

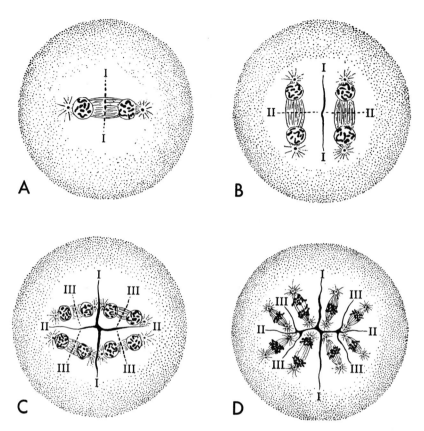

Fig. 20. Schematic diagrams of cleavage in the bird's egg, indicating the manner in which mitotic spindles become established with their long axis coinciding with the long axis of cytoplasmic mass. Following the division of the chromosomes, the cleavage furrow dividing the cytoplasm appears in the equatorial plane of the spindle, i.e., at right angles to the axis of the spindle. (*A*) First cleavage furrow appearing at dotted line I–I. (*B*) First cleavage furrow clearly marked and second appearing along dotted line II–II. (*C*) Third cleavage furrows forming. (*D*) Fourth cleavage spindles formed; additional cleavage furrows will appear later in positions indicated in Fig. 19, *D.*

orientation of the cleavage planes is to cut them in an apple. Let the core of the apple represent the animal-vegetative axis of the egg. The first cleavage furrow can be represented by notching the apple lengthwise, that is, as one ordinarily starts to split an apple into halves. The second cleavage furrow can be represented by cutting into the apple again in a plane passing through the axis of the core, but at right angles to the first cut, as one would start to quarter the apple.

The third cleavage furrows are variable in number and in position. In the most typical cases each of the four blastomeres established by the first two cleavages divides again so that eight blastomeres are formed (Fig. 19, C). Frequently, however, the third cleavage appears at first in only two of the blastomeres so that six cells result instead of eight.

The fourth series of cleavages takes place in such a manner that the central (apical) ends of the eight cells established by the third cleavage are cut off from their peripheral portions. The combined contour of the fourth cleavage furrows forms a small, irregularly circular furrow, the center of which is the point at which the first two cleavage planes intersect (Fig. 19, D). The central cells now appear completely separated in a surface view of the blastoderm, but sections show them still unseparated from the underlying yolk.

After the fourth, the succession of cleavages becomes irregular. In surface view it is possible to make out cleavage furrows that divide off additional "apical" cells (i.e., cells located centrally in the blastoderm and consequently at the apical or animal pole of the egg) and other, radial furrows that further divide the peripheral cells. Figures 19, E and F, show the increase in number of cells and their extension out over the surface of the yolk, resulting from the succession of cleavages. When the process of segmentation has progressed to the stage in which the cleavage planes are irregularly placed and the number of cells considerable, the term blastoderm is applied to the entire group of blastomeres formed by the segmentation of the blastodisk.[1]

In addition to the cleavages which are indicated on the surface, vertical sections of the 32-cell embryo show cleavage planes of an entirely different character. These cleavages appear below the surface and parallel to it. They establish a superficial layer of cells which are completely delimited. These superficial cells rest upon a layer of cells which are continuous on their deep faces with the yolk. Continued divisions of the same type eventually establish several strata of superficial cells. This process appears first in the central portion of the blastoderm. It progresses centrifugally as the blastoderm increases in size but does not extend to its extreme margin. The peripheral

[1] While but a single spermatozoön takes part in fertilization, other spermatozoa become lodged in the cytoplasm of the blastodisk. The nuclei of these spermatozoa migrate to the peripheral part of the blastoderm where they are recognizable for some time as the so-called accessory sperm nuclei. Some of them appear to undergo divisions which are accompanied by slight indications of division in the adjacent cytoplasm. The short superficial grooves thus formed are termed accessory cleavage furrows. No cells are formed by the accessory "cleavages." The sperm nuclei soon degenerate, the superficial furrows fade out, and usually as early as the 32-cell stage all traces of the process have disappeared without, as far as is known, affecting in any way the development of the embryo.

margin of the blastoderm remains a single cell in thickness, and the cells there lie unseparated from the yolk.

In looking at the arrangement of cells during cleavage and the formation of the morula, we are likely to pass all too unthinkingly over the complicated mechanisms that are involved. During the cleavage divisions, although the total protoplasmic mass of the embryo does not increase appreciably, there is a marked increase in the amount of chromosomal material. Were this not so, each succeeding cell generation would be poorer in chromatin than its predecessor. This augmentation of the DNA-containing material between successive cell divisions is termed *replication*. It is a most amazing process, for it involves far more than the mere synthesis of an exceedingly complex chemical substance. This new chromosomal material must duplicate the genetic pattern set up in the zygote at the time of fertilization. The dividing cells of an embryo must pass on not only the means of mere increase in the mass of the body, but also the inherited potentialities which make it possible for the embryo to differentiate a whole complement of organs like those of its parents. There is also the synthesis of considerable amounts of ribonucleic acid (RNA). Ribonucleic acid is sometimes referred to as a "messenger substance" because it is believed to carry "instructions" from the DNA of the chromosomes to the cytoplasm, directing its activities. These activities include all phases of the metabolic processes involved in the formation of additional cytoplasm including the synthesis of the necessary enzymes. Without the formation of new materials of these types, the growth potentialities of the small cell cluster that is the young embryo would soon be exhausted. We cannot see these processes as we look at young embryos, but we should realize that they are going on, and that the unraveling of their intricacies is one of the most challenging problems of embryology.

CHAPTER   4

ESTABLISHING THE GERM LAYERS AND FORMATION OF THE PRIMITIVE STREAK

The Morula Stage   It should by no means be inferred that cell division occurs less actively when the cleavage divisions have been accomplished. On the contrary the end of the "cleavage stage" is not marked by even a retardation in the succession of mitoses. As a matter of convenience in description, cleavage as a phase of development is regarded as ending when the progress of events ceases to be indicated merely by increase in the number of cells and begins to involve the localized aggregation and differentiation of various groups of cells. The nomenclature and limitation of the various periods of development are largely arbitrary, and the use of terms designating embryological stages should not be allowed to obscure the fact that the whole process is continuous and progresses from phase to phase without abrupt change or interruption. When we study conditions at what we are pleased to call "a stage of development" we are doing something comparable to looking with particular attention at one frame cut from a moving picture film. The frame we elect to scrutinize is selected because it catches something of interest in the progress of events, just as the advertizing pictures in front of a moving picture theater select and enlarge crucial scenes from the continuous action of the film.

In eggs without a large amount of yolk, segmentation results in the formation of a rounded, closely packed mass of blastomeres (Fig. 17, *E*). This is known as a morula from its resemblance to the mulberry fruit which is in form much like the more familiar raspberry or blackberry. At the end of segmentation the chick embryo has arrived at a stage which corresponds with the morula stage of forms with less yolk. It consists of a disk-shaped mass of cells several strata in thickness (the blastoderm) lying closely applied to the yolk. In the center of the blastoderm the cells are smaller and completely defined; at the periphery the cells are flattened, larger in surface extent, and are not walled off from the yolk beneath. At first glance the essential similarity of the chick morula with the morula of forms with less yolk might not be apparent. If, however, one imagines lifting the chick blastomeres off the surface of the inert yolk sphere on top of which they are growing and reshaping them into a sphere, the basic likeness of the morula of a chick to the morula of a form which has much less yolk, such as the frog, at once becomes apparent (Fig. 21).

Formation of the Blastula   The morula stage is of short duration. Almost as soon as it is established there begin to be changes presaging the formation of the blastula. In forms with a minimal amount of yolk there is a rapid rearrangement of the blastomeres which constituted the solid cellular sphere of the morula so that they come to form a thin outer layer, enclosing a newly formed central cavity known as the *segmentation cavity* or *blastocoele* (Fig. 22, *A*). With the establishing of the blastocoele the embryo is said to have progressed from the morula to the blastula stage. In forms, such as the Amphibia, with a moderate amount of yolk, the blastocoele becomes eccentrically displaced toward the animal pole because of the large size of the yolk-laden cells at the vegetal pole (Fig. 22, *B*). In the eggs of birds, with their enormous yolk-mass, the blastocoele is reduced to a mere slit-like cavity between a thin discoidal cap of cells and the underlying yolk (Fig. 22, *C*). The discoidal cap of cells above the blastocoele is called the *blastoderm*. The marginal area of the blastoderm, in which the cells remain incompletely walled off from the yolk and closely adherent to it, is called the *zone of junction*.

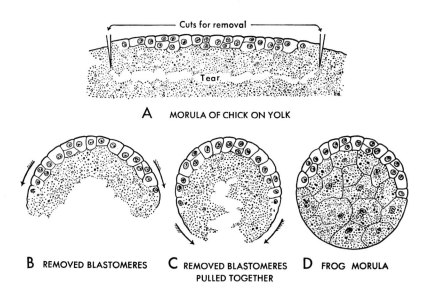

Fig. 21. Schematic diagrams comparing chick and amphibian morula stages. By removing the chick morula from the large yolk-sphere and pulling the margins together, its basic similarity to the amphibian morula is made evident.

In Fig. 22, *C*, only the blastoderm and the immediately underlying yolk are included in the diagram. At this magnification the complete yolk must be imagined as about 18 in. in diameter. The form of the chick blastula is readily understood if the full significance of the great yolk-mass is appreciated. Instead of being free to aggregate first into a solid sphere of cells and then into a hollow sphere of cells, as takes place in forms with little yolk, the blastomeres in the bird embryo are forced to grow on the surface of a large yolk sphere. Under such mechanical conditions the blastomeres are forced to become arranged in a disk-shaped mass on the surface of the yolk. If again at the blastula stage, as we did with the morula, we imagine the yolk removed and the disk of cells left free to assume a spherical shape, this time about a central cavity, its comparability with the blastula in a form having little yolk becomes apparent. There could be no such simple hollow sphere formation by rearrangement of the cells with the great bulk of the morula inert yolk. But the cells of the central region of the blastoderm are nevertheless separated from the yolk to form a small blastocoele with the yolk constituting its floor and at the same time, by reason of its great mass, nearly obliterating its lumen (Fig. 22, *C*).

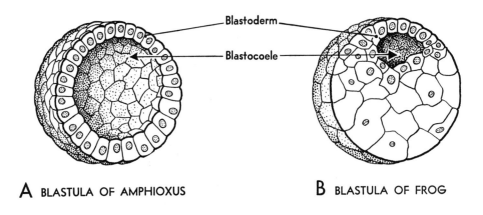

A BLASTULA OF AMPHIOXUS   B BLASTULA OF FROG

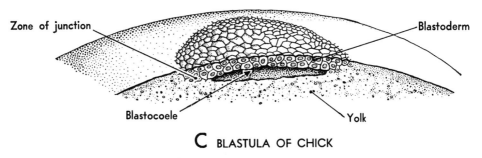

C BLASTULA OF CHICK

Fig. 22. Schematic sectional diagrams comparing the blastulae of Amphioxus, frog, and chick. (*Modified from Huettner, "Fundamentals of Comparative Embryology of the Vertebrates." By permission of The Macmillan Company, New York.*)

Gastrulation   The process of gastrulation begins as soon as blastulation is accomplished. With the appearance of this new process we think of the embryo as progressing into the next phase of its development, commonly called the "gastrula stage" (diminutive from Greek word *gaster* = stomach). In forms with little yolk, such as Amphioxus, gastrulation is essentially an impocketing of the blastula (Fig. 23, *A–D*). A double-layered cup is formed from a single-layered hollow sphere much as one might push in a hollow rubber ball with the thumb. The new cavity in the double-walled cup is termed the *gastrocoele* or *archenteron*. The opening from the outside into the gastrocoele has traditionally been called the *blastopore*, although it would be far more logical to call it the *gastropore*. Thus in gastrulation the single cell layer of the blastula has been differentiated to form two layers. The outer cell layer is known as the *ectoderm* and the inner layer as the *entoderm.* These germ layers, as they are called, differ from each other in their positional relationship to the embryo and to the surrounding environment. Each has different potentialities and each will in the course of development give rise to quite different types of tissues and organs.

The intriguing thing about these primary germ layers is not so much what we can see in them when they first arise, as the unseen developmental potentialities we know lie hidden within their cells. For we know that cells which look all alike to us when they first emerge as part of the ectoderm are destined to produce descendants as different as the horny outer cells of our skin, and the elaborate photosensitive rods and cones of our retina. Cells of the entoderm will become as unlike as the iodine-greedy cells of the thyroid, which produce the hormone thyroxin, and the ciliated epithelial cells moving the cleansing film of mucus over the lining of our respiratory passages. It is a realization of the divergent developmental routes to be followed by their cells that makes the emergence of the germ layers of so much interest. For the same reasons we now think of gastrulation more as the process which

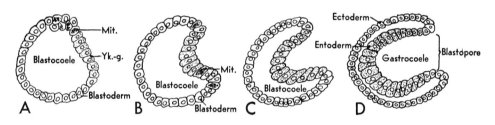

GASTRULATION IN FORM WITH ISOLECITHAL EGG HAVING ALMOST NO YOLK—AMPHIOXUS

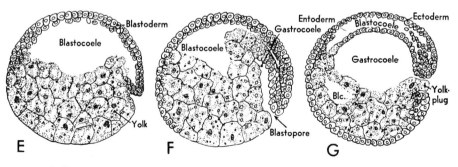

GASTRULATION IN FORM WITH TELOLECITHAL EGG CONTAINING MODERATE AMOUNT OF YOLK—AMPHIBIA

Fig. 23.   Schematic diagrams to show the effect of yolk on gastrulation.
   Abbreviations: Blc., blastocoele; *Mit.,* cell undergoing mitosis; *Yk.-g.,* yolk granules.

begins the segregation of the germ layers than as the mechanism of forming the gastrula cavity.

Effect of Yolk on Gastrulation   When there is an appreciable amount of yolk present in an egg we have already seen that it modified the configuration of the morula and blastula stages. It is only to be expected, therefore, that the process of gastrulation will be similarly affected. The most striking thing at first glance is, of course, the way the ingrowing cells are crowded away from the yolk-laden vegetal pole into the reduced blastocoele (Fig. 23, *E–G*). Less obvious, but of very great theoretical interest, are the differential growth rates correlated with the different extent to which yolk interferes with cell movements in different parts of the growing embryo. Some of the conditions seen at this stage of development in amphibian embryos help us to understand corresponding stages in chick development which we shall find even more extensively modified by reason of the far greater amount of yolk present in the hen's egg. For that reason it will be worth our while to examine in more detail the way amphibian embryos carry out the process of gastrulation.

Gastrulation in Amphibian Embryos   The general upward displacement of the process of gastrulation which is correlated with increase in the mass of the yolk present in different types of eggs has just been discussed. The yolk, also, radically affects the shape of the blastopore as can readily be seen if amphibian embryos are looked at, in caudal aspect, at various stages of gastrulation. In early stages the blastopore is a mere crescentic slit (Fig. 24, *A*), the region of inpocketing being crowded into such a restricted territory by the yolk. As the process gains momentum, the ingrowing margins, that is to say the *lips of the blastopore*, gradually are extended so they assume a circular shape as they turn inward around the yolk. This mass of yolk left presenting at the blastopore is known as the *yolk-plug* (Fig. 24, *B*, *C*). Finally differential growth crowds its lateral lips together and the closing blastopore becomes streak-like (Fig. 24, *D*, *E*).

With these matters of general orientation established we would like, now, to know more about the way the cells involved shift their position within the growing embryo, and the location of the areas of most rapid growth. It is possible by treating cells with nontoxic dyes such as Nile blue sulfate or neutral red to mark them and their immediate descendants. In this manner their changes of position can be followed through quite a period of growth before the dye becomes so much diffused that their identification is no longer possible. Marking may also be carried out by placing finely di-

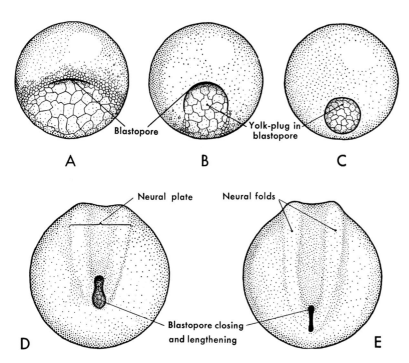

Fig. 24.   Caudal views of amphibian embryos, showing the changing configuration of the blastopore. (*Five stages selected and modified from Huettner, "Fundamentals of Comparative Embryology of the Vertebrates." By permission of The Macmillan Company, New York.*

vided, physiologically inert carbon particles, such as those of blood charcoal, on a small group of cells. By the patient following of embryos which have had strategically located cell groups marked by one or another of such methods, we have gradually learned much about the nature of the growth processes and cell movements involved in amphibian gastrulation. The process is by no means a simple matter of the infolding of a cell layer such as is usually understood by the term invagination.

Suppose, first, we follow the movements of a series of cells marked when they lie at the external surface of the embryo, peripheral to the blastopore, in locations such as those indicated by the black dots in Fig. 25, A. Such marked cells can be seen to move toward the margins of the blastopore in the directions indicated by the solid arrows. When they reach the lips of the blastopore, they turn inward and begin to move away from the blastopore (dotted arrows) as constituents of the newly established inner layers.

Marking experiments of this type also gives us the key to the elongation

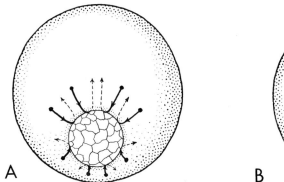
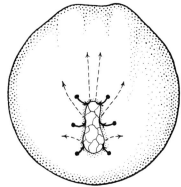

Fig. 25. Schematic diagrams indicating by arrows the way cells originally located in the primordial outer layer move first toward the blastopore (solid arrows) and then turn inward at the blastopore lips and move centrifugally (dotted arrows) in the newly established internal layer.

of the blastopore as it is closing. The inturning of cells at the upper margin of the blastopore, in other words at its so-called *dorsal lip*, is particularly active as indicated by the distance they travel in a given time. This is expressed graphically in Fig. 25, *B*, by the greater length of the dotted arrows near the dorsal midline. It is, of course, more or less figurative and definitely oversimplifying what actually happens, but we can think of the most dorsal part of the blastopore margin as being pulled on by the very active forward growth there going on. Concomitantly the convergence of superficial cells from either side appears to go on just a little more rapidly than the turned-in cells move away, so that the lateral margins of the blastopore tend to crowd toward the midline. Differential growth of this type is difficult to explain with static diagrams, but it may be most vividly demonstrated by time-lapse moving pictures. In this technique growing embryos are held quiet by anaesthetization and exposures made, every few seconds, over long periods of time. When such a film is projected at the usual rate of 16 to 24 frames per second, growth processes which move too slowly to be seen with the unaided eyes are sufficiently speeded up so that the directions of the movements involved, and their relative rates, can readily be seen. Because it is so difficult to describe succinctly, differential growth as a moulding factor in development is likely to be shunned in elementary treatises in a manner out of keeping with its importance.

By sectioning embryos at various periods after specifically located cell groups have been marked, it is possible to see more clearly what happens to cells from the different surface areas when they are carried to interior locations in the process of gastrulation. The sagittal section diagrams of Fig. 26 have been conventionally shaded to aid in following some of these movements. Around the margins of the blastopore and extending down onto the ventral part of the embryo is a region marked with bold black dots. This is spoken of as *prospective entoderm* because, if it is followed in the later stages, it can be seen to be rolled into the interior of the embryo and come to line its gastrocoele or primitive gut (Fig. 26, *B, C*).

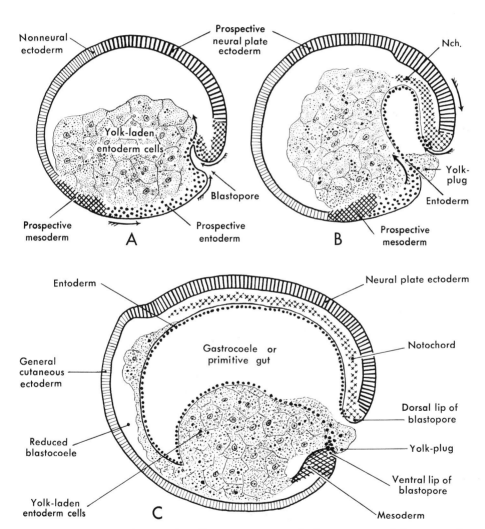

Fig. 26. Diagrams indicating the cell rearrangements which occur in amphibian gastrulation. [*Schematic sagittal sections based primarily on the work of Vogt, Arch. Entwickl.- mech. Organ.,* **120:** (1929).]

If, now, we look again at Fig. 26, *A*, there will be seen an area starting at the dorsal lip of the blastopore and extending on to the dorsal surface of the embryo which is marked with small crosses. This area is called *prospective notochord*, for when these cells move into the interior of the embryo in gastrulation they can be traced into the notochord (Fig. 26, *B*, *C*). In their changes in position these future notochordal cells have carried out two quite different moves in relation to the rest of the embryo. Take, for example, the most dorsally located cells in the group. First they move toward the blastopore, traveling as a part of the outer cell layer. When cells move thus over the spheroidal surface of a growing embryo, the process is spoken of as *epiboly*. At the same time that some of the prospective notochordal cells are moving toward the blastopore by epiboly, other cells that have already arrived there are rolling in around the lip of the blastopore, and moving away as part of a newly established internal structure. This process is called *involution*. It becomes obvious as soon as we thus begin to follow marked cells that gastrulation involves a certain amount of intucking of cells in its very early phases which comes within our usual meaning when we use the term invagination. It is equally obvious that, as the process continues, there is a very extensive amount of epiboly involved in bringing to the margin of the blastopore cells that are destined to roll around the blastopore lip by involution on their way to taking their places in an internal layer.

The origin of the mesoderm cannot be followed satisfactorily in sagittal sections alone. In such sections one sees from the first a small amount of *prospective mesoderm* at the dorsal lip of the blastopore (Fig. 26, *A–C*). Ventrally the prospective mesoderm is at first located quite a way from the lips of the blastopore (Fig. 26, *A*), but as the prospective entoderm moves along the outer surface toward the blastopore and then is carried in by involution, the prospective mesoderm follows it toward the blastopore until it, too, turns in (Fig. 26, *B*, *C*). What is not apparent from these diagrams based on sagittal sections is that there is really a ring of prospective mesoderm girdling the embryo. Reconstructions superimposing the mesoderm on shadowed sagittal sections show the way this zone of prospective mesoderm is carried back to the lips of the blastopore where the cells turn in between the involuting entoderm and the outermost cell layer of the embryo to spread out in a sheet-like layer investing the newly formed primitive gut (Fig. 27).

Even at the cost of considerable circumlocution we have been avoiding the use of the term ectoderm thus far in the discussion. It used to be customary, even at these early stages, to call whatever was on the outside of the embryo ectoderm. This led to the illogical terminology of having the ecto-

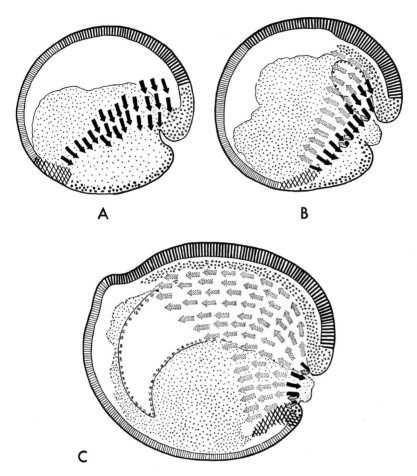

Fig. 27.   Diagrams showing the spread of the mesoderm in the embryos of tailed amphibia. These diagrams are built on the outlines of the same stages represented in Fig. 26. The solid arrows indicate the paths of cells moving toward the blastopore in the surface layer; the dotted arrows indicate their movement away from the blastopore after they have turned in and become incorporated in the invaginated layers. Note that the mesoderm is extended from the entire circumference of the lips of the blastopore, but that its growth from the dorsal lip is most vigorous. [*Modified from Vogt, Arch. Entwickl.-mech. Organ.,* **120** (1929).]

derm give rise to the entoderm and the ectoderm. Even worse, cells near the blastopore which were "ectoderm" when an embryo was examined at 9 A.M., had moved over the lip of the blastopore and become entoderm before noon. What is really happening, as we can see from studies with marked cells, is that a primitive layer, by cell rearrangement and specialization, is sorting itself out into cell groups which we are pleased to call ectoderm, entoderm, and mesoderm. If we use a nonspecific term like *protoderm*[1] for the original outer layer before the prospective entoderm and mesoderm have been segregated and rearranged in their characteristic internal relation, much confusion would be avoided. If we look again at the schematic diagrams of Fig. 26, this time with reference to the ectoderm, it should be apparent that what we are going to designate as ectoderm throughout all the later stages of development is that part of the protoderm which remains as an external layer after the entoderm, the notochord, and the rest of the mesoderm have been carried to their interior location in the process of gastrulation.

In addition to marking cells and watching where they move in the growing embryo, another most valuable method of analyzing the developmental potentialities of different areas is that of growing small cell groups taken out of their normal position in the embryo and seeing how they develop. Such cell clusters can be cultivated in glass chambers on artificial nutrient media made up to resemble the blood plasma of the species being studied. The method is spoken of technically as "cultivation of explants *in vitro.*" It offers both a check on, and a supplement to, marking experiments. By gathering together information obtained by these two types of studies it is possible to make a map of the location of the prospective regions for all of the germ layers while the cells that will later be segregated to compose them still lie in the outer layer of the young embryo just commencing to undergo gastrulation (Fig. 28).

Gastrulation in Birds   If the processes involved in the gastrulation of amphibia have been mastered, there should be no difficulty in understanding gastrulation in bird embryos. The same basic processes are involved. The only essential difference is that in birds the yolk has become so large that, instead of merely altering the shape and location of the blastopore, as it does in amphibia, it prevents the formation of any open blastopore at all. We can, nevertheless, unmistakably identify the area that is homologous

---

[1] Blastoderm implying as it does a rapidly growing layer would be an excellent term had it not been widely employed also in a less restricted sense. The term *protoderm* seems equally appropriate and has not been made equivocal by previous use in a different sense.

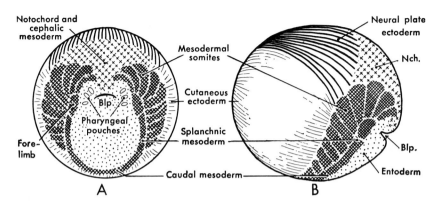

Fig. 28. Prospective areas of the embryos of tailed amphibians at the stage when gastrulation is just beginning. [*Modified from Vogt, Arch. Entwickl.-mech. Organ.*, **120** (1929).] (*A*) Caudal aspect; (*B*) lateral aspect.

Abbreviations: Blp., blastopore; *Nch.*, notochord.

with the blastopore and see a similar migration of cells toward it by epiboly, and a similar turning in of cells from the primordial outer layer to become entoderm and mesoderm.

We have already traced the establishing of the chick's blastula as a disk of cells lying on the yolk but separated from it centrally by a flattened blastocoele or segmentation cavity (Fig. 22, *C*). The peripheral part of the blastoderm where the cells lie unseparated from the yolk has been termed the area opaca because in preparations made by removing the blastoderm from the yolk surface, yolk adheres to it and renders it more opaque (Figs. 2 and 39, *C*). This opacity is especially apparent when a preparation is viewed under the microscope by transmitted light. The central area of the blastoderm, because it is separated from the yolk by the segmentation cavity, does not bring a mass of adherent yolk with it when the blastoderm is removed. It is for this reason translucent and is called the area pellucida.

The area opaca later becomes differentiated so that three more or less distinct zones may be distinguished: (1) a peripheral zone, known as the *margin of overgrowth*, where rapid proliferation has pushed the cells out over the yolk without their becoming adherent to it; (2) an intermediate zone, known as the *zone of junction*, in which the deep-lying cells do not have complete cell boundaries but constitute a syncytium blending without definite boundary into the superficial layer of white yolk and adhering to it by means of penetrating strands of cytoplasm; (3) an inner zone, known as the *germ wall*, made up of cells derived from the inner border of the zone of junction which have acquired definite boundaries and become more or less free from the yolk. The cells of the germ wall usually contain numerous small yolk granules which were enmeshed in their cytoplasm when they were, as cells of the zone of junction, unseparated from the yolk. The inner margin of the germ wall marks the transition from area opaca to area pellucida (Fig. 39, E).

The changes in the blastula which presage the beginning of gastrulation occur in the first few hours after the egg is laid. In an egg that has been incubated from 3 to 4 hours, one quadrant of the area pellucida is definitely thickened as compared with the others. This thickening marks the future caudal end of the embryo (Fig. 29, A). Two or three hours later the thickened area has become much more definitely outlined and is beginning to show some cephalocaudal elongation (Fig. 29, B). By 7 to 9 hours of incubation the

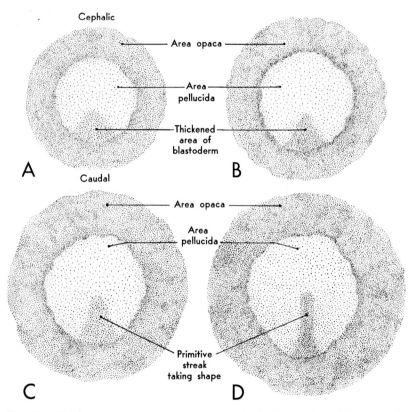

Fig. 29. Chick embryos showing four stages in the formation of the primitive streak. (A) 3 to 4 hours' incubation. (B) 5 to 6 hours' incubation. (C) 7 to 8 hours' incubation. (D) 10 to 12 hours' incubation. [*Based in part on the photomicrographs of Spratt, J. Exptl, Zool.,* **103** (1946).]

elongation is still more definite (Fig. 29, *C*), and by the end of the first half-day the thickened area has assumed a shape which has led to its being called the *primitive streak* (Fig. 29, *D*).

By about 16 hours of incubation the primitive streak has become so prominent that embryos are characterized as being in the primitive streak stage (Fig. 30). Closer examination of the primitive streak in fixed and stained whole-mounts shows it to be composed of a central furrow, the *primitive groove*, flanked on either side by thickened margins, the *primitive ridges* (Figs. 30 and 39, *D*). At the cephalic end of the primitive streak, closely packed cells form a local thickening known as *Hensen's node* (Fig. 30). The part of the area pellucida adjacent to the primitive streak is beginning to show increased thickness which in the next 2 hours becomes more clearly marked and is said to constitute the embryonal area (Fig. 38). Because of its shape the embryonal area is frequently spoken of as the *embryonic shield*. The process of elongation that the primitive streak has undergone in its formation (cf. Figs. 29, *B–D* and 30) is shared by the area pellucida. Instead of its early, almost circular outline the area pellucida in an embryo with a fully formed primitive streak has become elliptical. The long axis of the primitive streak definitely establishes the long axis of the future embryonic body. The caudal end of the streak is the one which lies close to the area opaca. The cephalic end of the streak lies well within the boundary between area pellucida and opaca and also is marked by Hensen's node.

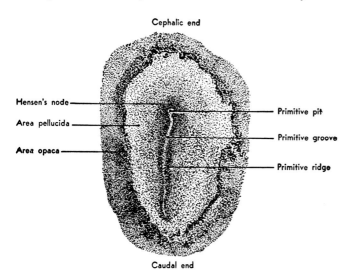

Fig. 30. Dorsal view (×14) of entire chick embryo in the primitive-streak stage (about 16 hours of incubation).

Having become familiar with the nomenclature and topography of embryos in the primitive streak stage, we are ready to look more closely at the nature of the growth processes involved in this phase of development. One of the most instructive studies on the nature of the cell movements involved is that of Spratt (1946). His method was to place a series of carbon spots on the blastoderms of embryos just as they were about to form the primitive streak and then grow them on plasma clots so the marked areas could be followed. The results of hundreds of such experiments he has summarized in the form of a graphic table are reproduced here as Fig. 31. Each row of diagrams should be followed from left to right. The left-hand diagram indicates where the ink marks were made in relation to the thickened caudal quadrant of the preprimitive streak stage (cf. Fig. 29, *A*). The diagrams in the center show where the marked spots had moved when the partially formed primitive streak was still short (cf. Fig. 29, *B*). In the right-hand column the diagrams show where the marked areas had moved by the time a definitely elongated primitive streak had been formed (cf. Fig. 29, *C*). Systematic and thoughtful study of these series of markings will be found most instructive. From rows I and II it will be apparent that cells in the caudal quadrant are moving mesially into the region where the streak is taking shape. It is highly significant that certain cells from the surface layer are moving toward the primitive streak of the chick just as they moved in amphibian embryos toward the blastopore. Row III shows the way the growing streak extends to include more and more of the superficially located cells lying just cephalic to it, in the midline. Rows IV and V show that cells in the blastoderm situated far laterally or cephalically do not become incorporated in the streak at all, but move peripherally with the growth of the blastoderm.

It becomes more difficult to trace their movements with precision as the cells become concentrated in the thickened part of the blastoderm that is being reshaped to form the primitive streak. It seems reasonably clear from Spratt's marking experiments, however, that following their migration from outlying surface areas many of the cells sink to a deep-lying position in the thickened zone and thence move inwards to take part in the formation of the entoderm (Fig. 33, *A*). It is apparent, also, that other cells move from the inner part of the surface cell layer directly into the expanding layer of entoderm cells by a process characterized as *delamination* (Figs. 32 and 33, *A*).

In the light of the foregoing facts it seems reasonable to regard the pre-primitive streak thickened area of the chick blastoderm as the symbolic homolog of a blastopore that could not open because of the impeding effect of the enormous yolk-mass. As has been mentioned, the movement of surface cells toward this thickened area is strongly suggestive of the way cells move by epiboly toward the amphibian blastopore. The emergence of cells from this thickened zone to enter the entoderm is as near as the chick, with its impeding yolk-mass, can come to duplicating the involution of entoderm at the lips of the blastopore as it occurs in amphibian embryos (cf. Figs. 26, *A* and 33, *A*). The only differences seem to be the lack of an open blasto-pore[1] in the chick, and the way many of the cells split directly off the under side of the outer layer as if, to speak figuratively, they were impatient of the impediment offered by the yolk and the closed blastopore, and were taking a "short cut" to their appointed place in the entoderm.

If we follow the process of gastrulation in bird embryos into its later phases when notochord and mesoderm are being established, the fact that the primitive streak represents fused blastopore lips becomes even more apparent. Shortly after the primitive streak has been formed and the ento-derm is well established, cells begin to push in from the region of Hensen's node to form the rod-like notochord in the midline beneath the ectoderm (Fig. 33, *B*). The area where the chick notochord is thus formed clearly cor-responds with the dorsal lip of the blastopore where the amphibian noto-chord arises (Fig. 26). Sections taken across the primitive streak caudal to Hensen's node show the mesoderm extending out on either side between the ectoderm and the entoderm (Fig. 33, *C*). The relationships are, again, essen-tially the same as those seen in amphibian embryos with the mesoderm arising from the fused lips of the blastopore and extending between ecto-derm and entoderm (Figs. 26 and 27).

Potencies and Prospective Areas in Chicks at the Primitive Streak Stage
The use of vital dyes and carbon particles to mark specific cell areas for subsequent tracing of their positional changes is already familiar. Another exceedingly valuable method of determining what specialized role a partic-ular cluster of undifferentiated cells is going to play in future developmental processes is to excise it, and grow it outside the embryonic body. After it has been allowed sufficient time to express its inherent developmental po-tentialities, such a piece is sectioned and studied microscopically to ascertain the kind of tissue into which it has developed. A convenient and widely used

---

[1] It used to be believed that the chick had, for a brief time, an open slit-like blastopore from which the first entoderm cells were involuted in a manner closely similar to gastrulation as it occurs in amphibia. Because of its appealing consistency from the standpoint of comparative embryology, this interpretation was particularly tenacious in textbooks, including the previous editions of this book. All recent critical studies indicate, however, that the yolk so strongly modifies gastrulation that, as presented in this re-vised edition, no open blastopore is formed. This correction, however, in no way changes the basic con-cept that the primitive streak of the chick represents the fused lips of the blastopore of forms with less yolk.

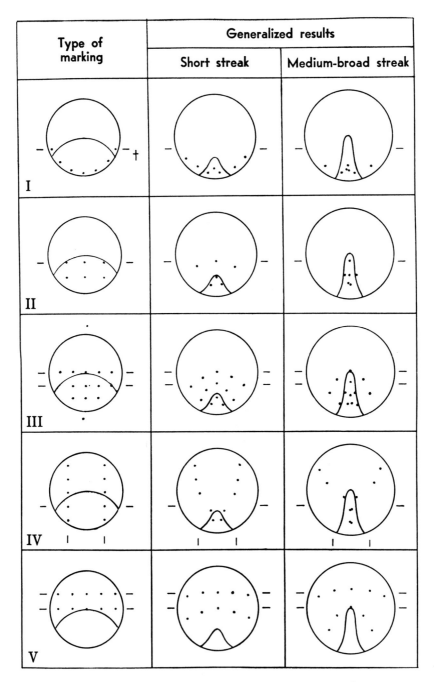

Fig. 31. Graphic summary of a series of experiments to show cell movements in the neighborhood of the primitive streak during its formation. Cells of living

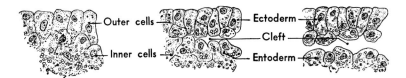

Fig. 32. Delamination of the cells of the early blastoderm (protoderm) to establish separate ectodermal and entodermal layers. [*Redrawn, with some modifications, from Peter, Z. Mikr. Anat. Forsch.,* **43** (1938).]

method of growing such explants from chick embryos is to graft them into the chorio-allantoic membrane of an older chick embryo, say one of 9 or 10 days. This highly vascular tissue close beneath the porous shell makes an excellent culture medium where the explant can grow uninfluenced by the proximity of the tissues which would have lain next to it in its normal location in the body, and might have influenced its development. This method of chorio-allantoic grafting has been used to test small pieces cut from all parts of embryos in the primitive streak stage. When, for example, a small cluster of ectoderm cells from a region a little to one side of, and slightly cephalic to, Hensen's node is permitted to develop as a chorio-allantoic graft, it forms cells, some of which have unmistakably the histological peculiarities of retinal elements. We say the cells in question exhibit a *potency* for eye formation. If an area where a particular potency has been located is explored in more detail, it is found that there is a certain central part of it from which practically all the explants exhibit the potency in question. Explants taken farther peripherally show the potency in decreasing percentages of the grafts made. The territory where a specific potency is regularly manifested is said to be the *presumptive center* for the organ in question.

Another example of work of this type pertains to the identification of cells with heart-forming potencies. Rudnick has shown that explants from the mesoderm adjacent to Hensen's node on either side (Fig. 35) have the potentiality of developing into cardiac tissue which unmistakably declares itself by developing contractile activity. More recently, by a series of cell

---

embryos were marked in the locations indicated by the dots in the diagrams in the left-hand column. The embryos were then cultured, and the marked cells moved into the position indicated by the dots in the diagrams of the middle and right-hand columns. The marginal horizontal lines in the left-hand column are placed at the initial level of some of the spots marked. Carrying these lines at this same level in the right-hand columns helps following the growth changes involved. [*From Spratt, J. Exptl. Zool.,* **103** (1946).]

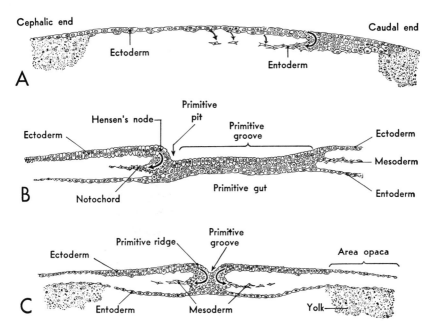

Fig. 33. Schematic diagrams indicating some of the cell movements involved in the gastrulation of chick embryos. (*A*) Longitudinal section of the blastoderm from a pre-primitive streak chick during entoderm formation. The origin of some entoderm cells by delamination is suggested by the cells indicated by small arrows. (*B*) Longitudinal plan of embryo of approximately 17 hours of incubation to show the relations of the growing notochord, Hensen's node, and the primitive pit. (*C*) Cross section of embryo in the primitive-streak stage to show the turning in of cells at the primitive groove to enter the mesodermal layers.

marking experiments carried out with magnificent technical skill, De Haan and Rosenquist have been able to identify cells destined to come into these heart-forming areas even before they moved from the protoderm through the primitive streak to take their place in the mesoderm.

Maps of the prospective areas for chicks of the primitive streak stage are reproduced as Figs. 34 and 35. It should be emphasized that the sharpness of the boundaries between different prospective areas as shown in such figures is an entirely artificial device for vivid graphic presentation. In actuality there are vague transition zones rather than anything like the sharp delimitations of the schematic diagram. Many interesting forecasts of events to come, developmentally speaking, can be gleaned from the study of such diagrams of prospective areas. Perhaps, however, for the beginner their greatest value will be realized by referring back to them as older embryos are studied which show the emergence of the structures here represented only by prospective areas and presumptive centers.

Embryological Importance of the Germ Layers   In looking back over the development thus far undergone by the embryo, perhaps the most conspicuous thing, at first glance, is the multitude of cells formed from the single fertilized egg cell. Of more significance, however, is the fact that even during the early phases of rapid proliferation the cells thus formed do not remain as an unorganized mass. Almost at once they become definitely arranged on the yolk sphere as the cap-like blastoderm. Scarcely is the blastoderm established when, at a definite thickened area, certain cells loosen themselves and become arranged as a second layer beneath the first. This second layer because of its position inside the original layer is called the entoderm. Shortly a third cell layer makes its appearance between the first two, being called, appropriately enough, the mesoderm. That part of the original cap of cells which still constitutes the outer layer after the entoderm and mesoderm have been established is now properly called the ectoderm.[1] These three cell layers are spoken of as the *germ layers* of the embryo.

The germ layers are of interest to the embryologist from several points of view. The simple organization of the embryo when it consists of first a single, then two, and finally three primary structural layers is reminiscent of ancestral adult conditions that occur in primitive groups of the invertebrate series. From the standpoint of probable ontogenetic recapitulations of remote phylogenetic history, several facts are quite suggestive. The nervous system of the vertebrate embryo arises from the ectoderm—the layer through which a primitive organism which has not as yet evolved a central nervous system is in touch with its environment. The lining of the vertebrate digestive tube is formed from the entoderm—the layer which in very primitive forms lines a gastrocoele-like enteric cavity. The vertebrate skeletal and circulatory structures are derived almost entirely from the mesoderm—the layer which in small, lowly organized invertebrates is relatively inconspicuous, but which constitutes a progressively greater proportion of the total bulk of animals as they increase in size and complexity and consequently need more elaborate supporting and transporting systems.

Interesting as are the possibilities of interpreting the germ layers from

---

[1] British authors generally use instead of ectoderm, mesoderm, and entoderm—epiblast, mesoblast, and hypoblast. These terms have the disadvantage of not implying the layer-like character expressed by the root *derm* (skin), but they have the advantage of emphasizing the actively growing character of the layers by the incorporation of the Greek root *blast*, meaning primarily a sprout or shoot of a plant, but extended in biological usage to designate anything that is growing rapidly.

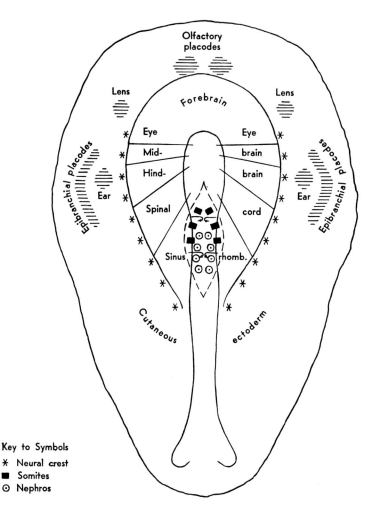

Fig. 34. Map of the prospective areas of the outer layer of the chick embryo in the primitive-streak stage. [*Modified from Rudnick, Quart. Rev. Biol.,* **19** (1944).]

the standpoint of their phylogenetic significance, our chief present concern with them centers about the part they play in the development of the individual. The establishment of the germ layers is the first segregation of cell groups which are clearly distinct from one another by reason of their definite relations within the embryo. The fact that these relations are fundamentally the same in all vertebrate embryos speaks forcefully of the common ancestry and similar heritage of the various members of this great group of animals.

It means, furthermore, that in these germ layers we have a common starting point in the fabrication of the variations which different classes of animals have built upon the common underlying plan of body structure characteristic of the vertebrate group as a whole.

The establishment of the germ layers marks also a transition from the

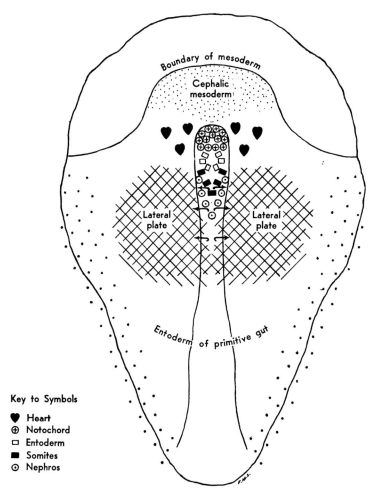

Fig. 35. Map of the prospective areas of the invaginated layers of the chick in the primitive-streak stage. The symbols for entoderm appearing toward the cephalic part of the primitive streak are meant to indicate that some prospective entoderm cells are still moving in through the streak. As shown in Fig. 33 the entoderm has already spread out widely to form the roof of the primitive gut. [*Modified from Rudnick, Quart. Rev. Biol.,* **19** (1944).]

period of development when mere increase in number of cells is the outstanding event, to one in which differentiation and specialization are dominating concomitants of growth. Differentiation is occurring within the germ layers even before we can see tangible evidences of it by any of our present microscopical methods. Within a layer that looks all alike to us there are gradually being established groups of cells with different developmental potentialities. Subtle changes precede or accompany this regrouping of cells. With succeeding divisions the future fate of a cell group becomes increasingly sharply fixed along certain lines. After a time it can be altered by experimental interference only within a very narrow range. This process which establishes the kind of end result that may be expected from a given cell group is spoken of as *determination*. It is mediated through gene control, operating in the immediate environment furnished by the growing body of the embryo.

The time at which determination occurs can be ascertained only by such experimental procedures as transplanting apparently undifferentiated cells, or growing them in tissue cultures and watching them declare their potentialities. The bodily segregation of cell groups that follows or accompanies their determination is, by contrast, easy to follow in the ordinary sections prepared for microscopic study. Some of the various ways it may be accomplished are suggested in the diagrams of Fig. 36. When a group of cells is outpocketed from a parent layer, the process is spoken of as *evagination*. If the segregation is accomplished by inpocketing, we call it *invagination*. Particularly in the mesoderm, *migration* of cells may be an essential preliminary to their *aggregation* in a new area and their subsequent *differentiation* into specialized structures.

From primordial cell aggregations established by such processes the organs with which we are familiar in the adult gradually take shape. Thus at some time during the course of its development each region of the young embryonic body loses its initially diverse potentialities and becomes subdivided into parts, each limited to a specific type of differentiation. Such limitation of the fate of a cell group is called its *determination*. The story of the embryological origin of the various parts of the body is, therefore, the history of the growth, subdivision, and differentiation of the germ layers with the determination of one specific area after another within them. This sequence of events is so dramatic, and seemingly so irrevocable if one studies only normal development, that a word of caution is needed. Transplantation experiments not only permit us to map presumptive areas, but they show us, also, that presumptive areas grafted into an abnormal location may show considerable modification of what appeared to be their predetermined fate.

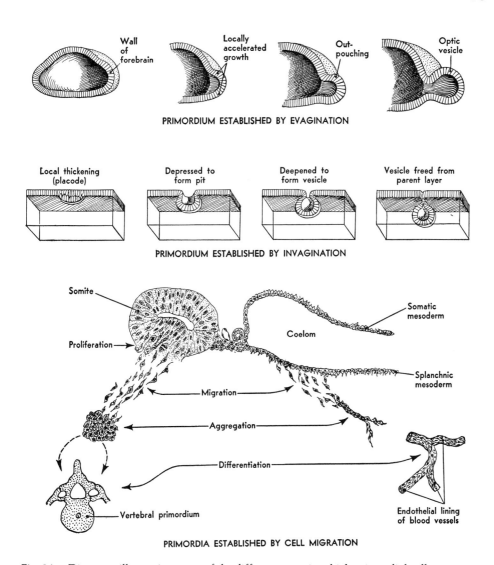

Fig. 36. Diagrams illustrating some of the different ways in which primordial cell groups may arise from parent cell layers.

There must be the clear realization that to attain its normal destiny a group of embryonic cells must have not only the necessary inherent developmental potentialities but also the proper environment.

An outline plan of the repeated regroupings and progressive differentiations and specializations of the cells derived from the primary germ layers is given in Fig. 37. This chart at the present stage of our study will serve as

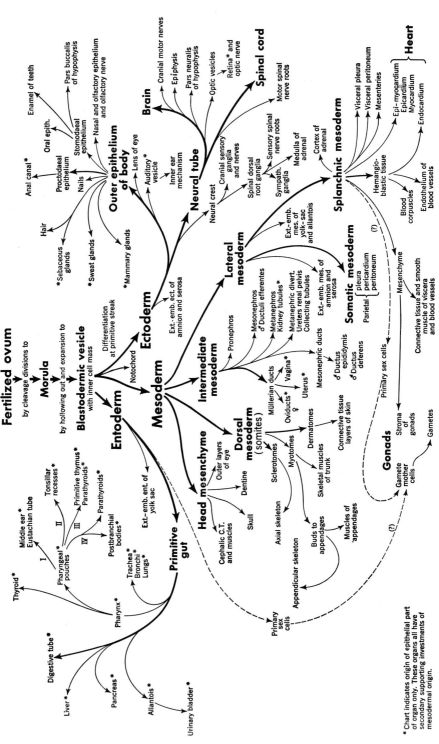

Fig. 37. Chart showing derivation of various parts of the body by progressive differentiation and divergent specialization. Note especially how the origin of all the organs can be traced back to the three primary germ layers.

*Chart indicates origin of epithelial part of organ only. These organs all have secondary supporting investments of mesoderm origin.

a means of pointing out in a general way whither the early processes with which we have been dealing are destined to lead. As we follow the phenomena of development farther, we shall find each natural division of the subject centers more or less sharply about some particular branch of this genealogical tree of the germ layers.

CHAPTER 5

FROM THE PRIMITIVE STREAK STAGE TO THE APPEARANCE OF THE SOMITES

The Notochord   Those who have studied comparative anatomy will be familiar with the notochord as one of the characteristic structural features of all vertebrates. Its presence reveals the relationship of forms so primitive —or so retrogressed—that they do not show any typical vertebral column. In fishes such as those of the shark family which have only a cartilaginous skeleton, the notochord is clearly recognizable in the adult as an axial structure about which the ring-like cartilage vertebrae have developed. When, in the progress of evolution, cartilaginous vertebrae are replaced by highly developed bony vertebrae, the notochord becomes much reduced. Even in mammals, however, a minute canal persists in the center of the developing vertebrae marking the position of the notochord. In young embryos of birds or mammals the notochord is for some time a conspicuous structure, an enduring record of evolutionary history, and at the same time an advance indication of the location where the vertebral column will later be formed.

Starting as soon as the primitive streak was fully established, we saw that the ingrowth of cells from the cephalic part of the primitive streak about Hensen's node initiated the formation of the notochord (Fig. 33, *B*). In chick embryos that have been incubated about 18 hours, the notochord has become markedly elongated to form a distinct structure, extending cephalad in the midline from Hensen's node (Fig. 38). This origin of the notochord from cells at the cephalic end of the primitive streak, just as in other more primitive vertebrate embryos it arises from the region of the dorsal lip of the blastopore, is one of the most cogent of the points confirming the interpretation of the chick primitive streak as representing the fused blastopore lips of forms with less yolk.

In the older embryological treatises the young notochord was frequently called the "head process." Following this usage chick embryos of about 18 hours incubation, when the notochord is one of the few prominent structural features they possess, are often spoken of as being in the "head process stage." It is not a particularly fortunate term but it is one that is so firmly entrenched that it must be understood if one is to do any delving into the literature.

Sections of embryos at this stage of development (Fig. 39) give much

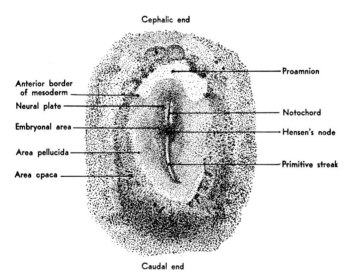

Cephalic end

Anterior border of mesoderm

Neural plate

Embryonal area

Area pellucida

Area opaca

Proamnion

Notochord

Hensen's node

Primitive streak

Caudal end

Fig. 38. Dorsal view (×14) of entire chick embryo of 18 hours' incubation.

more information on the structure and relations of the notochord than can be gleaned from the study of whole-mounts. In its more cephalic portions it is a fairly sharply circumscribed structure (Fig. 39, *A*). Farther caudally it is made up of diffusely arranged cells merging with other layers at Hensen's node (Fig. 39, *F*). The experimental studies of Spratt (1947) indicate that the notochord is molded *in situ* out of previously involuted cells in a manner comparable to the way we shall see the somites molded from the segmental zone of the mesoderm in slightly older embryos.

In the sections diagrammed in Fig. 39 a different conventional scheme of representation has been employed to indicate each of the germ layers. The ectoderm is vertically hatched, the cells of the mesoderm are represented by heavy angular dots when they are isolated or by solid black lines when they lie arranged in the form of compact layers, and the entoderm is represented by fine stippling backed by a single line as in Fig. 39, or by coarse dots backed by a single line as in Fig. 51. A similar conventional representation of the different germ layers is observed in all diagrams of sections in order to facilitate following the way in which the organ systems of the embryo are constructed from the germ layers. Details of cell structure have been shown for only a few regions where they are of particular significance. It is hoped that these few examples will encourage the student to observe cell structure in his own study of sections. The plane in which each of the sections diagrammed passes through the embryo is indicated by a line drawn on a small outline sketch of an embryo of corresponding stage. For interpretation these outline sketches should be compared with detailed drawings of entire embryos of the same stage of development, or better, with actual specimens.

There has been considerable disagreement among embryologists as to how to describe the origin of the notochord in terms of germ layers. Most of the difficulties in this regard could be obviated by the use of the term protoderm, as previously suggested, for the original outer layer of the embryo. If this is done we can simply recognize the fact that prospective notochordal tissue is rolled in at the blastopore, or its homolog, in a manner similar to the way the prospective mesoderm is segregated from the protoderm. This makes it logical to regard notochordal tissue as merely a special part of the general mesoderm which in turn fits in nicely with the way the notochord and certain parts of the mesoderm both form supporting structures. For these reasons the notochord in this text is regarded as a special part of the mesoderm and so represented in the scheme of conventionalizing the different germ layers in the diagrams.

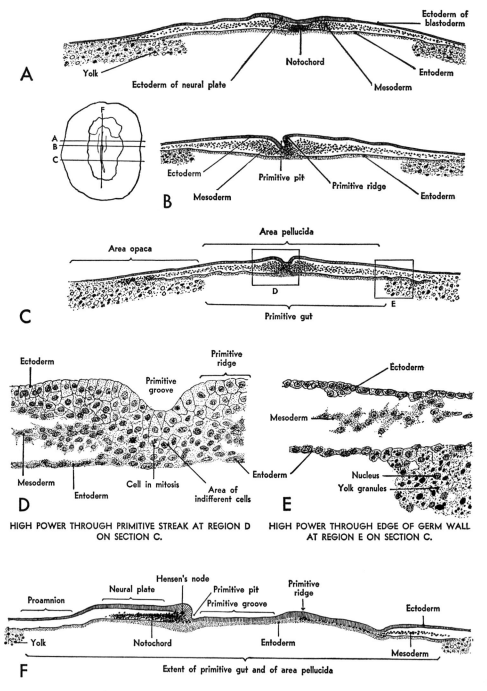

A

Yolk   Ectoderm of neural plate   Notochord   Ectoderm of blastoderm   Mesoderm   Entoderm

B

Ectoderm   Primitive pit   Primitive ridge   Mesoderm   Entoderm

C

Area opaca   Area pellucida   D   E   Primitive gut

D

Ectoderm   Primitive groove   Primitive ridge   Mesoderm   Entoderm   Cell in mitosis   Area of indifferent cells

HIGH POWER THROUGH PRIMITIVE STREAK AT REGION D
ON SECTION C.

E

Ectoderm   Mesoderm   Entoderm   Nucleus   Yolk granules

HIGH POWER THROUGH EDGE OF GERM WALL
AT REGION E ON SECTION C.

F

Proamnion   Neural plate   Hensen's node   Primitive pit   Primitive groove   Primitive ridge   Ectoderm   Yolk   Notochord   Entoderm   Mesoderm

Extent of primitive gut and of area pellucida

Fig. 39.   Sections of 18-hour chick. The location of each section is indicated by a line drawn

The Primitive Streak as a Point of Reference in Early Growth   As was pointed out in the previous chapter, the primitive streak is of great interest from the standpoint of comparative embryology because it represents the fused lips of the blastopore of more primitive forms which are endowed with less yolk. From the standpoint of the ontogenetic development of the chick, the primitive streak demanded special attention because of the way it was involved in the initial segregation of the germ layers. Another feature to which we should now turn our attention is the growth activity of the cells in the primitive streak region itself and of those cells that have recently moved into adjacent internal layers by way of the primitive streak. By their active proliferation all the germ layers of the embryo continue to expand.

We have just been considering the growth of the notochord from the cephalic end of the primitive streak, which is of course particularly well shown in sagittal sections (Fig. 39, *F*). Cross sections passing through the primitive streak a little caudal to Hensen's node are most suitable for studying the relations of the rest of the mesoderm to ectoderm and entoderm (Fig. 39, *C*). In the floor of the primitive groove the ectoderm dips down toward the entoderm and in this region there is no line of demarcation between the two layers, both of them merging in a mass of cells that can best be characterized as indifferent (Fig. 39, *D*). Emerging on either side from this indifferent cell area is the mesoderm, growing laterad into the space between ectoderm and entoderm. If the movements of cells in the early stages of primitive streak formation are recalled (review Figs. 31 and 33, *C*), it is apparent that the floor of the early primitive groove must contain a succession of protoderm cells that have been converging there to move thence into the inner layers. These migrating cells must pass through the "indifferent area" on their way to take their appointed places. During all these cell movements there is of course concomitant cell division, so that this area remains, during all the early stages of development, a territory from which cells are passed on to enter the differentiated germ layers.

---

on a small outline sketch of an entire embryo of corresponding age. The letters affixed to the lines indicating the location of the sections correspond with the letters designating the section diagrams. Each germ layer is represented by a different conventional scheme: ectoderm by vertical hatching; entoderm by fine stippling backed by a single line; the cells of the mesoderm, which at this stage do not form a coherent layer, by heavy angular dots. (*A*) Diagram of transverse section through neural plate and notochord. (*B*) Diagram of transverse section through primitive pit. (*C*) Diagram of transverse section through primitive streak. (*D*) Drawing showing cellular structure in primitive-streak region. (*E*) Drawing showing cellular structure at inner margin of germ wall. (*F*) Diagram of median longitudinal section passing through notochord and primitive streak.

In Figure 39, *D*, a small region at the primitive streak has been drawn at higher magnification to show the characteristic cellular structure of the undifferentiated region in the floor of the primitive groove and of the various layers continuous with one another at this place. The cells of the ectoderm are much more closely packed together and more sharply delimited than those of the other germ layers. Where the ectoderm is thickened in the primitive ridge region (Fig. 39, *D*), it has become several cell layers in depth (stratified). In regions lateral to the primitive ridge it gradually becomes thinner until it consists of but a single cell layer (Fig. 39, *E*). The rapid extension that the mesoderm is at this time undergoing is indicated by the loose arrangement and sprawling appearance of its cells. Their irregular cytoplasmic processes make them look much like amoebae fixed during locomotion. The cells of the entoderm are neither as closely packed nor as clearly defined as are the ectoderm cells. Nevertheless, in contrast to the condition of the mesoderm at this stage, the entoderm cells form a definite, unbroken layer.

Regression of the Primitive Streak   Following the phases of its greatest conspicuousness in young embryos the primitive streak rapidly becomes both relatively and actually smaller. There is still proliferative activity in and around it, but the cells move into other areas, and the streak itself is shortened. Concurrently the rapid growth of the body cephalic to the primitive streak makes it appear to be shifted to a relatively more caudal position in the embryonic body. The marking experiments of Spratt again give us the clearest picture of what is occurring. Carbon spots placed in the superficial layer close to the primitive streak (Fig. 40, row I) indicate that, even after the streak has begun to shorten, there is still some convergence of cells toward it, although this is far less active and extensive than it was in earlier stages. Marks placed at the two ends of the streak (row II) retain their relative positions within the streak and approach each other as the streak shortens. This is clearly indicative that the rapid growth of the embryo in length during the period the primitive streak is shortening is not, as has been widely believed, due to contributions from the cephalic end of the primitive streak but rather to growth in the embryonic areas in front of the streak. Marking of the tip of the young notochord (row III) confirms this point. By marking a central spot, as well as the two ends of the streak (row IV), it can be established from the way the middle mark approaches the caudal one that most of the shortening of the streak is due to the emigration of cells from its caudal half. These cells spread out laterally and caudally contributing to the expansion of both the extra-embryonic and intra-embryonic layers. Finally the striking increase in the length of the embryos that accom-

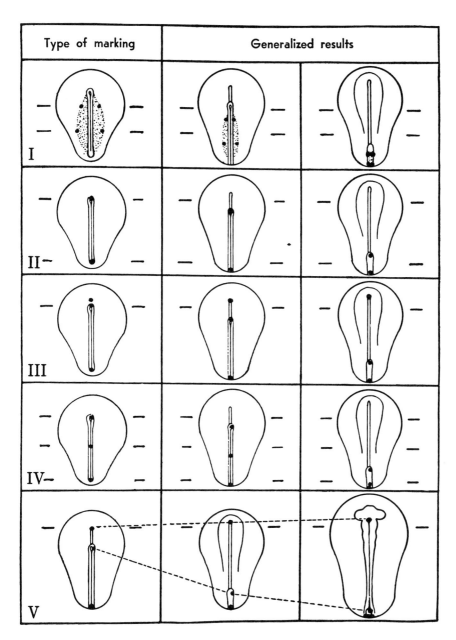

Fig. 40. Diagrams indicating the movements of marked cells in the neighborhood of the primitive streak and Hensen's node. The position of the marked cells at the beginning of the experiment is indicated by the dots in the diagrams in the left-hand column. The diagrams in the center and right-hand columns indicate the locations to which these marked cells moved in embryos kept alive by tissue-culture techniques. [*From Spratt, J. Exptl. Zool.,* **104** (1947).]

panies the diminution of the primitive streak is vividly shown by again marking the cephalic tip of the young notochord (row V) and this time following its growth into considerably older stages. The major points in these changing relations are summarized quantitatively in Fig. 41.

As the primitive streak thus undergoes a sort of dismemberment from its more caudal portion, Hensen's node moves farther and farther back toward the area opaca. By the end of the second day of incubation little more than the node and a very short portion of the streak remains. At this stage what remains of the once prominent node and the primitive streak is usually given a new name, the *end-bud*.

Caudal Growth and Cephalic Precocity   The very active growth in the region just ahead of the primitive streak which we saw clearly indicated by the rapidly elongating notochord continues long after the primitive streak has undergone regression. One is likely, in looking casually at a series of embryos in which the progress of elongation in the more anterior regions is so striking, to attribute it to particularly active growth in this region itself. In reality it is due rather to rapid growth from behind which pushes the cephalic region ahead of it. The fact that the growth of a young embryo is taking place chiefly from its caudal end has a bearing on the relative progress

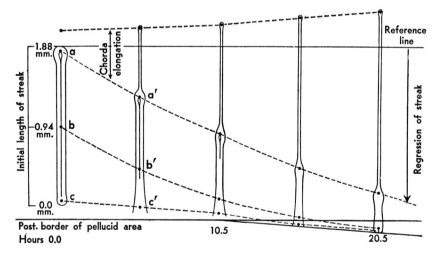

Fig. 41.   Graphic summary of the regression of the primitive streak and the growth of the notochord. The scale is indicated by the length of the primitive streak given in millimeters at the left. The time is given in hours beyond the start of the experiments. [*From Spratt, J. Exptl. Zool.*, **104** (1947).]

of differentiation in different regions of the body. It is a striking fact that the cephalic end of an embryo will always be found precocious in differentiation as compared with the more posterior portions of the embryo. This much-commented-on condition seems but natural when we consider that the head is actually older in development; for the structures posterior to the head are laid down by cells which were proliferated from the caudally located areas of rapid growth subsequently to the establishment of the head itself. Differentiation does, of course, occur exceedingly rapidly in the head. Were this not so, other regions would pass it in developmental progress. But we cannot, in taking cognizance of this condition, afford to overlook the fact that the head is given a considerable lead at the outset by its earlier establishment.

Growth of the Entoderm and Establishing of the Primitive Gut   Sections of embryos which had been incubated from 18 to 20 hours showed how the entoderm has spread out and become organized into a coherent layer of cells merging peripherally with the inner margin of the germ wall and overlapping it to a certain extent (Fig. 39, C, E, F). The cavity between the yolk and the entoderm which has been called the gastrocoele is now termed the *primitive gut.* The yolk floor of the primitive gut does not show in sections prepared by the usual methods. The reasons for this are to be found in the relations of the embryo to the yolk before it is removed for sectioning. In the entire central region of the blastoderm the yolk is separated from the entoderm by the cavity of the primitive gut. When the embryo is removed from the yolk sphere, the yolk floor of the primitive gut, not being adherent to the blastoderm, is left behind (Fig. 42, A). In contrast, the peripheral part of the blastoderm lies closely applied to the yolk. Some yolk adheres to this part of the blastoderm when it is removed. This adherent yolk is shown in the section diagrams of Fig. 39. Its presence clearly indicates why this region (area opaca) appears less tranlucent in surface views of entire embryos (Fig. 38). This spread out arrangement of the developing walls of gut is so unlike conditions seen in adults or in embryos of forms with less yolk that it sometimes bothers students encountering it for the first time. It might be of assistance in getting these relations in proper perspective to picture a chick lifted off the yolk and the lateral margins pulled together ventrally (Fig. 42). If one chooses, the method of comparison may readily be reversed by imagining an amphibian embryo split open along its midventral line and spread out on the surface of a sphere as a chick lies on the yolk.

In embryos of 18 hours the primitive gut is a cavity with a flat roof of entoderm and a floor of yolk. Peripherally it is bounded on all sides by

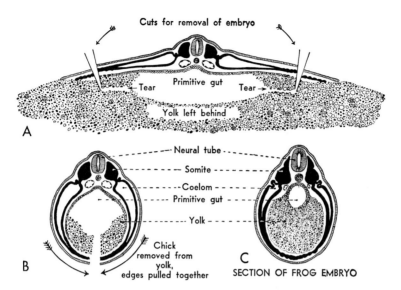

Fig. 42. *(A)* is a diagram showing how the usual method of removing chick embryos from the yolk in order to prepare them for microscopical study makes the sections appear as if the primitive gut had no ventral boundary. *(B)* and *(C)* show how removing a chick from the yolk and pulling its edges together ventrally facilitates comparisons with forms which do not develop with their growing bodies spread out on a large yolk-sphere.

the germ wall (Fig. 39, *C, F*). The merging of the cells of the entoderm with the yolk-mass is shown in the small area of the germ wall drawn to a high magnification in Fig. 39, *E*. In the germ wall cell boundaries are incomplete and very difficult to distinguish, but nuclei can be made out surrounded by more or less definite areas of cytoplasm. This cytoplasm contains numerous yolk-granules in various stages of absorption. It will be recalled that the nuclei of the germ wall arise by division from the nuclei of cells lying at the margins of the expanding blastoderm. They appear to be concerned in breaking up the yolk in advance of the entoderm as it is spreading about the yolk sphere.

By about the 20th hour of incubation indications can be seen of a local differentiation of that region of the primitive gut which underlies the anterior part of the embryo. By focusing through the ectoderm in the anterior region of a whole-mount of this age, a pocket of entoderm can be seen (Fig. 43). This entodermal pocket is the first part of the gut to acquire a floor, other than the yolk floor, and is called from its anterior position the *foregut*. Consideration of the foregut except to note the location of its first appearance can advantageously be deferred because its origin and relationships are more readily appreciated from the study of somewhat older embryos.

Growth and Early Differentiation of the Mesoderm   The mesoderm which arises from either side of the primitive streak spreads rapidly laterad, and at the same time each lateral wing of the mesoderm swings cephalad. Figure 44, *A–C* shows schematically the extension of the mesoderm during the latter part of the first day of incubation. The diagonal hatching represents the mesoderm seen through a supposedly transparent ectoderm. The principal landmarks of the embryos are sketchily represented.

It will be noticed that the manner in which the mesoderm spreads out leaves a mesoderm-free area in the anterior portion of the blastoderm. This region, known as the *proamnion*, is clearly recongizable in entire embryos by reason of its lesser density (Figs. 43 and 45). The name is unfortunate because of its false implication that this region is in some way a precursor of the amnion. Although it has long been known to be a misnomer, the term proamnion is so deeply entrenched in the literature that it is difficult to dislodge. The only importance of this area is the information its rapidly di-

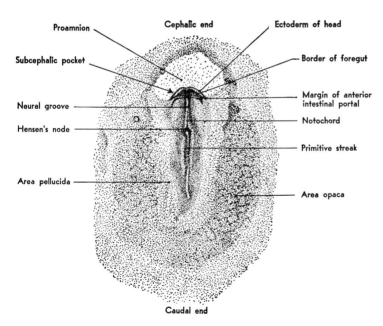

Fig. 43. Dorsal view ($\times 14$) of entire chick embryo of about 20 hours' incubation.

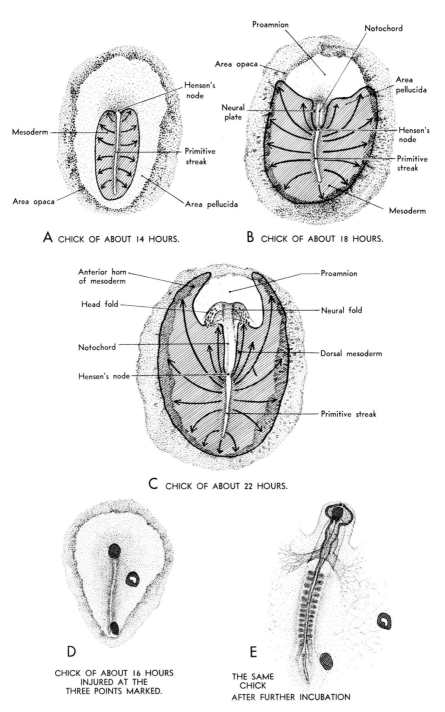

A CHICK OF ABOUT 14 HOURS.

B CHICK OF ABOUT 18 HOURS.

C CHICK OF ABOUT 22 HOURS.

D

CHICK OF ABOUT 16 HOURS
INJURED AT THE
THREE POINTS MARKED.

E

THE SAME
CHICK
AFTER FURTHER INCUBATION

Fig. 44. Diagrams showing direction of growth from the primitive streak as a

minishing size gives as to the rate of growth of the mesodermic wings which form its lateral borders (Fig. 43, *A* to *C*).

As the notochord is more definitely molded the rest of the mesoderm tends to pull away from it somewhat so that, especially in the more cephalic regions, the notochord is left sharply outlined in the midline (Figs. 44, *C*, and 50 *A*, *B*) in territory for a time relatively free of other mesoderm. Sections passing through the primitive streak of embryos of about 18 hours' incubation age show loosely aggregated masses of mesoderm extending to either side between the ectoderm and entoderm. Immediately adjacent to the primitive streak the mesoderm is markedly thicker than it is farther laterally (Fig. 39, *B*, *C*). This tendency for the mesoderm to be thicker near the midline extends cephalad from the level of the primitive streak to involve the mesoderm adjacent to the neural tube and the notochord (Fig. 39, *A*). These zones of thickened mesoderm either side of the midline are commonly referred to as *dorsal mesoderm* because of the position they are destined to occupy when the body is closed in ventrally. (Look ahead at Fig. 112.) Because of the way in which they become divided into the metamerically arranged masses called somites, they are frequently designated, also, as the *segmental zones* of the mesoderm. The segmental zones are in early stages most clearly marked somewhat cephalic to Hensen's node, where the first somites will appear. In the embryo represented in Fig. 45 there is just a suggestion of the clefts separating the first somites. As the segmental zones of the mesoderm are followed from this region caudad on either side of the primitive streak, they gradually become less and less definite. Traced laterally, the mesoderm rapidly becomes less thick. This thinner region at the sides contains the primordial tissue from which the *intermediate zones* and the *lateral plates of the mesoderm* are later differentiated.

The rapid extension that the mesoderm is at this time undergoing is indicated by the loose arrangement and sprawling appearance of its cells. Their irregular cytoplasmic processes make them look much like amebas fixed during locomotion. As a matter of fact they were actually on the move before they were killed by fixation. Between the germ layers of an embryo,

---

center. (*A*) to (*C*) show the growth as expressed by the progress of the mesoderm during the latter part of the first day of incubation. Some of the more prominent structural features of the embryos are drawn in lightly for orientation but the ectoderm is supposed to be nearly transparent allowing the mesoderm to show through. The areas into which the mesoderm has grown are indicated by diagonal hatching. (*D*) and (*E*) show the direction of growth as demonstrated by experimental methods. (*After Kopsch.*) (*D*) shows the location at which three injuries were made close to the primitive streak of a 16-hour embryo. (*E*) shows the position to which the injured areas were carried by growth of the same embryo subsequent to the operation.

what looks like empty space in a section was occupied in the living embryo by a noncellular, somewhat gelatinous material that can best be termed *interlaminar jelly*. This serves as a substratum in which many of the young mesodermal cells move by their own activity. When a cell thus breaks loose from the coherent mesodermal layer and starts to move through the inter-laminar jelly to some other place where it will settle down and become dif-ferentiated, we speak of it as a *mesenchymal cell*. An aggregation of mesen-chymal cells supported in interlaminar jelly is called *mesenchyme*.

The mesoderm of the cephalic region is of the mesenchymal type. It is derived primarily from cells which become detached from the more com-pact layers of mesoderm lying farther caudally in the body and migrate into the cephalic region. Recent studies suggest that, in certain areas, these cells are supplemented by others emerging from the neural tube and becoming unidentifiably mingled with cells of mesodermal origin. Whatever their origin may be, because of their positional relations to ectoderm and ento-derm and because of their wandering proclivities, all the cells of the middle layer in the cephalic region are best designated as mesenchymal.

Formation of the Neural Plate   In surface views of entire chicks of about 18 hours (Fig. 38) areas of greater density may be made out on either side of the notochord. These areas extend somewhat rostral to the cephalic end of the notochord, where they appear to blend with each other in the midline. Sections of this region (Fig. 39, *A*) show that the greater density seen in whole-mounts is due to thickening of the ectoderm. Rapid cell proliferation has resulted in the ectoderm in the middle region becoming several cells in thickness. This thickened area is known as the *neural (medullary) plate*. Laterally the thickened ectoderm of the neural plate blends without abrupt transition into the thinner ectoderm of the general blastodermic surface. Rostrally the neural plate is more clearly marked than it is caudally. At the level of Hensen's node the neural plate diverges into two elongated areas of thickening, one on either side of the primitive streak.

In embryos of about 22 hours (Fig. 45) the neural plate becomes longi-tudinally folded to establish a trough known as the *neural groove*. The bot-tom of the neural groove lies in the middorsal line. Flanking the neural groove on each side is a longitudinal ridge-like elevation involving the lateral portion of the neural plate. These two elevations which bound the neural groove laterally are known as the *neural folds*. The folding of the originally flat neural plate to form a gutter, flanked on either side by parallel ridges, is an expression of the same extremely rapid cell proliferation which first manifested itself in the local thickening of the ectoderm to form

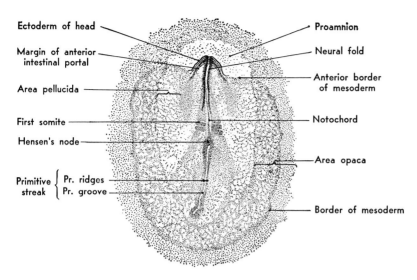

Ectoderm of head

Margin of anterior intestinal portal

Area pellucida

First somite

Hensen's node

Primitive { Pr. ridges
streak   { Pr. groove

Proamnion

Neural fold

Anterior border of mesoderm

Notochord

Area opaca

Border of mesoderm

Fig. 45. Dorsal view ($\times$14) of an entire chick embryo at the beginning of somite formation. (About 22 to 23 hours of incubation.)

the neural plate. The formation of the neural plate and its subsequent folding to form the neural groove are the first indications of the differentiation of the central nervous system.

Experimental studies of the formation of the neural plate have yielded extraordinarily interesting information as to the way one part of a developing embryo may influence the differentiation of other parts. When this occurs it is spoken of as *induction*. The tissue which calls forth the response is called the *inductor*, or *organizer*, and the chemical substance it gives off is known as the *organizing substance* or *evocator*. Induction has been most extensively studied in amphibian embryos and it was in such forms that the primary organizing effect of the chorda-mesoderm on the development of the neural plate was first discovered by Spemann and Mangold. These classical experiments and others that followed showed that if contact of chorda-mesoderm tissue with the overlying ectoderm is prevented, no neural plate is formed. Even more dramatic are the complementary experiments in which the chorda-mesoderm is transplanted so it underlies an area of ectoderm not in the dorsal midline. Responding to the presence of the organizer, this ectoderm, which does not normally do anything of the kind, is induced to differentiate into neural plate tissue.

It is highly significant that organizing effects may be produced by other than living tissue. Chorda tissue, killed by boiling, will still call forth the potentiality of the ectoderm to form a neural plate. So also will small pieces of agar that have been soaked in the juices from crushed organizing tissue. Such experiments clearly indicate that the action depends on the formation within the cells of the organizer of a chemical substance which when released calls forth the characteristic response in another tissue. These evocators, however, are not all powerful in development, for only specific types of tissue will respond to them. Thus, although neural plate formation can be induced in areas of the ectoderm where it does not normally occur, it cannot be induced at all from entoderm or mesoderm. This again emphasizes that, as transplantation experiments so clearly showed, the various growing tissues of an embryo have certain inherent developmental potentialities of their own. These may be activated to further development or, within limits, have the course of their development altered from the usual pattern by the action of an organizer, but the necessary potentialities must be there before any response is possible. When a tissue is thus able to respond to an organizing substance by producing a definite specialized tissue it is said to possess *competence* in this regard.

It is of interest, also, that an organizing substance obtained from one species may be effective when introduced into the embryo of a different species. Waddington and Schmidt (1933) grafted notochordal tissue from a duck into a chick embryo of the primitive streak stage, placing it slightly to one side of the midline. After being permitted to grow for a time following this operation, the embryo was fixed and sectioned. Over the grafted duck notochord the chick ectoderm had responded by forming a second neural tube alongside of its own normally developing and normally placed neural tube (Fig. 46). The same experiment was also successfully carried out in the reverse direction with the duck as host and the chick as donor. Extending this work to less closely related species, Waddington (1934) succeeded in inducing neural plate tissue in a rabbit embryo by grafting in potential chorda tissue from the cephalic end of a chick primitive streak. Work of this type is relatively new and technically so difficult and tedious that results come slowly. Much remains to be learned, but the few experiments cited as examples will suffice to show the immense vistas that are currently being opened by studies of organizers and the responses of embryonic tissues to them.

Differentiation of the Embryonal Area   Due to the thickening of the ectoderm to form the neural plate and also to the thickening of the dorsal zones of the mesoderm, the part of the blastoderm immediately surrounding the primitive streak and notochord has become noticeably more dense than that

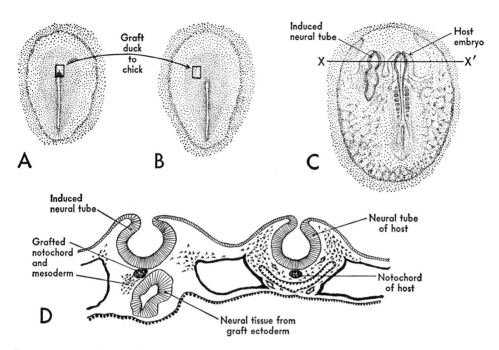

Fig. 46. Semischematic drawings showing the induction of an accessory neural tube as a result of grafting notochordal tissue from a duck donor into a chick host. [*Based on the work of Waddington and Schmitt, Arch. Entwickl.-mech. d. Organ.*, **128** (1933).] (*A*) Duck embryo showing the location from which the graft was taken. (*B*) Chick host showing the location where the graft was implanted. (*C*) Embryo cultivated for 31½ hours after implanting of the graft, showing the location of the induced accessory neural tube. (*D*) Section at level of the line X-X' in *C*, diagrammed with the same conventional representation of the component layers as that employed in the other sectional diagrams in this text.

in the peripheral portion of the area pellucida. Because it is the region in which the embryo itself is developed, this denser region is known as the *embryonal area*. Although the embryonal area is at this early stage directly continuous with the peripheral part of the blastoderm without any definite line of demarcation, they later become folded off from each other. The peripheral portion of the blastoderm is then spoken of as extra-embryonic because it gives rise to structures which are not built into the body of the embryo, although they play a vital part in its nutrition and protection during development.

The anterior region of the embryonal area is thickened and protrudes above the general surface of the surrounding blastoderm as a rounded elevation. This prominence marks the region in which the head of the embryo will develop (Figs. 43 and 45). The crescentic fold which bounds it is termed the *head fold* and is the first definite boundary of the body of the embryo. Throughout the course of development we shall find the cephalic region farther advanced in differentiation than other parts of the body. The importance of differential growth as a factor in establishing this condition ontogenetically has already been discussed. Those familiar with comparative anatomy will see in this precocity of the head a repetition of race history in the development of the individual, for phylogenetically the head is the oldest and most highly differentiated region of the body. This cephalic precocity of the embryo is but one of many manifestations of the law of recapitulation, in conformity with which the individual in its development rapidly repeats the main steps in the development of the race to which it belongs.

CHAPTER 6

STRUCTURE OF TWENTY-FOUR-HOUR CHICKS

Formation of the Head   In embryos of about 22 hours we saw that the anterior part of the embryonal area had been thickened and elevated above the level of the surrounding blastoderm, with a well-defined crescentic fold marking its anterior boundary (Fig. 45). In the next 3 or 4 hours the cephalic region undergoes rapid growth. Its elevation above the blastoderm becomes much more marked, and it extends anteriorly, so it overhangs the proamnion region (Fig. 47). The crescentic fold which formerly marked its anterior boundary appears to have undercut the anterior part of the embryo and separated it from the blastoderm. The changes in relationships are due, however, not so much to a posterior movement of the fold as to the anterior growth of the embryo itself. This anterior region which projects, free from the blastoderm, may now properly be termed the head of the embryo. The space formed between the head and the blastoderm is called the *subcephalic space*, or *pocket* (Fig. 50, *E*).

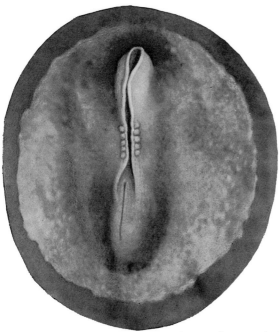

Fig. 47. Chick embryo of 25 to 26 hours photographed by reflected light to show its external configuration. Compare with Figs. 48 and 49.

In the midline the notochord can be seen through the overlying ectoderm. It is larger posteriorly near its point of origin than it is anteriorly. Nevertheless it can readily be traced into the cephalic region where it will be seen to terminate somewhat short of the rostal end of the head (Fig. 48).

Formation of the Neural Groove   The neural plate in chicks of 18 hours was seen as a flat, thickened area of the ectoderm. In embryos of 21 to 22 hours a longitudinal folding had involved it, establishing the neural groove in the middorsal line flanked on either side by the neural folds. At 24 hours of incubation the folding of the neural plate is much more clearly marked. In transparent preparations of the entire embryo (Fig. 48) the neural folds appear as a pair of dark bands. The folding which establishes the neural groove takes place first in the cephalic region of the embryo. At its cephalic end the neural groove is therefore deeper, and the neural folds are correspondingly more prominent than they are caudally. The folding has not, at this stage, been carried much beyond the cephalic half of the embryo. Consequently as the neural folds are followed caudad, they diverge slightly from each other and become less and less distinct.

Study of transverse sections of an embryo of this stage affords a clearer interpretation of the conditions in neural groove formation than the study of entire embryos. A section passing through the head region (Fig. 50, *A*) shows the neural plate folded so it forms a nearly complete tube. Dorsally the margins of the neural folds of either side have approached each other and lie almost in contact. The formation of the neural folds takes place first in about the center of the head region and progresses thence rostrad and caudad. By following caudad the sections of a transverse series, the margins of the neural folds will be seen less and less closely approximated to each other.

Establishing the Foregut   In the outgrowth of the head, the entoderm as well as the ectoderm has been involved. As a result the entoderm forms a pocket within the ectoderm, much like a small glove finger within a larger. This entodermic pocket, or foregut, is the first part of the digestive tract to acquire a definite cellular floor. That part of the gut caudal to the foregut, where the yolk still constitutes the only floor, is termed the *midgut*. The

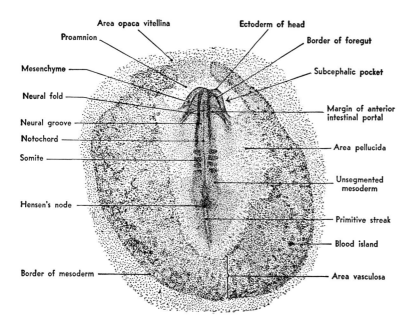

Fig. 48.   Dorsal view (×16) of entire chick embryo having four pairs of mesodermic somites (about 24 hours' incubation). Compare this figure of an embryo, which has been stained and cleared and then drawn by transmitted light, with the preceding figure which shows an embryo of about the same age photographed by reflected light.

opening from the midgut into the foregut is called the *anterior intestinal portal* (Fig. 50, *E*).

The topography of the foregut region at this stage can be made out very well by studying the ventral aspect of entire embryos. The margin of the anterior intestinal portal appears as a well-defined crescentic line (Fig. 49). The lateral boundaries of the foregut can be seen to join the caudally directed tips of the crescentic margin of the portal. Considerably cephalic to the intestinal portal an irregularly recurved line can be made out. On either side it appears to merge with the ectoderm of the head. This line marks the extent to which the head is free from the blastoderm. It is due to the fold at the bottom of the subcephalic pocket, where the ectoderm of the under surface of the head is continuous with the ectoderm of the blastoderm. Comparison of Fig. 49 with the sagittal section diagrammed in Fig. 50, *E*, will aid in making clear the relationships of foregut to the head. From the sagittal section it will also be apparent why the margins of the intestinal portal and of the subcephalic pocket appear as dark lines in the whole-mount.

In viewing an entire embryo under the microscope by transmitted light one depends largely on differences in density for locating deep-lying structures. When a layer is folded so the light must pass through it edgewise, the fold stands out as a dark line by reason of the greater thickness it presents. (Study Figs. 2 and 3 systematically and thoughtfully.) Such a correlative study of whole-mounts and sections is essential in acquiring a three-dimensional concept of the relationships so important in understanding the

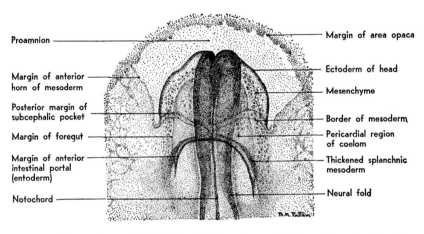

Proamnion

Margin of anterior horn of mesoderm

Posterior margin of subcephalic pocket

Margin of foregut

Margin of anterior intestinal portal (entoderm)

Notochord

Margin of area opaca

Ectoderm of head

Mesenchyme

Border of mesoderm

Pericardial region of coelom

Thickened splanchnic mesoderm

Neural fold

Fig. 49. Ventral view (×40) of cephalic region of chick embryo having five pairs of somites (about 25 to 26 hours of incubation).

structure of young embryos. Students working with serial sections for the first time frequently find this difficult to do. As an aid to the mastering of such an approach a stereogram of a transected 24-hour chick (Fig. 51) has been placed on a page facing the labeled diagrams of sections (Fig. 50).

Regional Divisions of the Mesoderm   The first conspicuous metamerically arranged structures to appear in the chick are the *mesodermic somites*. The somites arise by division of the mesoderm of the dorsal or segmental zone to form block-like cell masses. In the embryo shown in Fig. 48 three pairs of somites are completely delimited, and a fourth pair can be made out which is not as yet completely cut off from the dorsal mesoderm posterior to it. The regular addition of somites as embryos increase in age makes the number of somites a useful criterion of the stage of development. For this reason it is customary to designate the stage of development of young embryos on the basis of the number of somites they have formed. The word "pairs" is usually omitted, but it is understood that a 6-somite embryo is one with 6 pairs of somites.

Cross sections passing through the midbody region show the formation of the somites and the beginning of other changes in the mesoderm (Fig. 50, C, cf. also Fig. 81). Following the mesoderm from the midline toward either side, three regions or zones can be made out: (1) the dorsal mesoderm, which at this level has been organized into somites, (2) the intermediate mesoderm, a thin and narrow plate of cells connecting the dorsal and lateral mesoderm, and (3) the lateral mesoderm, which is distinguished from the intermediate by being split into two layers.

The somites are compact cell masses lying immediately lateral to the neural folds. The cells composing them have a fairly definite radial arrangement about a central cavity which is very minute or wanting altogether when the somites are first formed, but which later becomes enlarged (Fig. 81). Cephalic and caudal to the region in which somites have been formed, the dorsal mesoderm is differentiated from the rest of the mesoderm simply by its greater thickness and compactness.

In 24-hour embryos the *intermediate mesoderm* shows very little differentiation. In the chick it never becomes segmentally divided as does the dorsal mesoderm. The fact that it is potentially segmental in character is indicated, however, by the way in which it later gives rise to serially arranged nephric tubules. Because of the part it plays in the establishment of the excretory system the intermediate mesoderm is frequently called the *nephrotomic plate*.

In the chick the *lateral mesoderm*, like the intermediate mesoderm,

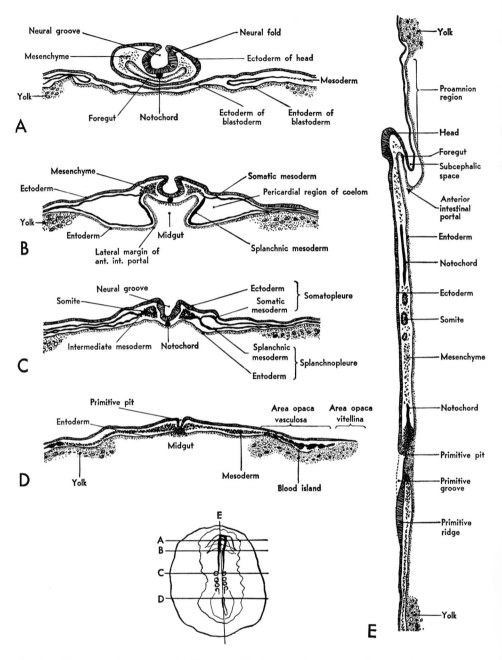

Fig. 50. Diagrams of sections of 24-hour chick. The levels of the sections are located on an outline sketch of the entire embryo. The conventional representation of the germ layers is the same as that employed in Fig. 39 except that here, where its cells have become aggregated to form definite layers, the mesoderm is represented by solid black lines.

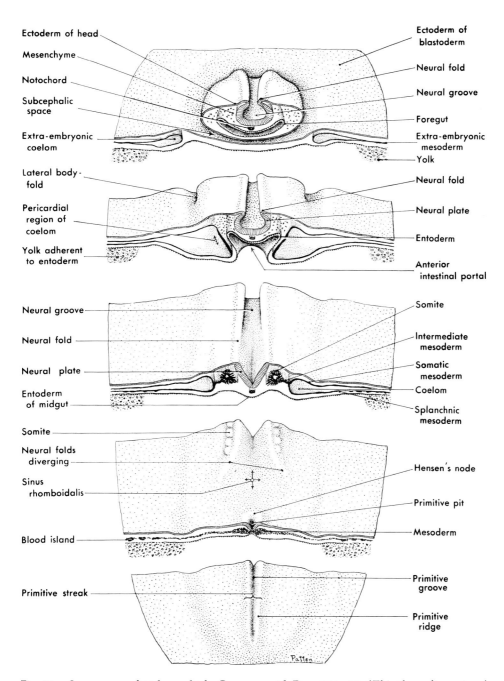

Ectoderm of head

Mesenchyme

Notochord

Subcephalic space

Extra-embryonic coelom

Lateral body-fold

Pericardial region of coelom

Yolk adherent to entoderm

Neural groove

Neural fold

Neural plate

Entoderm of midgut

Somite

Neural folds diverging

Sinus rhomboidalis

Blood island

Primitive streak

Ectoderm of blastoderm

Neural fold

Neural groove

Foregut

Extra-embryonic mesoderm

Yolk

Neural fold

Neural plate

Entoderm

Anterior intestinal portal

Somite

Intermediate mesoderm

Somatic mesoderm

Coelom

Splanchnic mesoderm

Hensen's node

Primitive pit

Mesoderm

Primitive groove

Primitive ridge

Patten

Fig. 51.   Stereogram of 24-hour chick. Compare with Figs. 47 to 50. (*This three-dimensional approach was suggested by Huettner's drawings in his "Fundamentals of Comparative Embryology of the Vertebrates," The Macmillan Company, New York.*)

shows no segmental division. In 24-hour embryos (Fig. 50, C) it is clearly differentiated from the intermediate mesoderm by being split horizontally into two layers with a space between them. The layer of lateral mesoderm lying next to the ectoderm is termed the *somatic mesoderm,* the layer next to the entoderm is termed the *splanchnic mesoderm,* and the cavity between somatic and splanchnic mesoderm is the *coelom.* Because in development the somatic mesoderm and ectoderm are closely associated and undergo many foldings in common, it is convenient to designate the two layers together by the single term *somatopleure.* Similarly the splanchnic mesoderm and the entoderm together are designated as the *splanchnopleure.*

The Coelom   The coelom, like the cell layers of the blastoderm, extends over the yolk peripherally beyond the embryonal area (Fig. 50, C). Later in development foldings mark off the embryonic from the extra-embryonic portion of the germ layers. This same folding process divides the coelom into intra-embryonic and extra-embryonic regions. In the 24-hour chick, however, embryonic and extra-embryonic coelom have not been separated.

It is evident from the manner in which the coelomic chambers arise in the lateral mesoderm that the coelom of the embryo consists of a pair of bilaterally symmetrical chambers. It is not until later in development that the right and left coelomic chambers become confluent ventrally to form an unpaired body cavity such as is found in adult vertebrates.

The Pericardial Region   In the region of the anterior intestinal portal the coelomic chambers on either side show very marked local enlargement. Later in development these dilated regions are extended mesiad and break through into each other ventral to the foregut to form the pericardial cavity. In their early condition these enlarged regions of the coelomic chambers are sometimes called amniocardiac vesicles. With their later fate in mind we may avoid multiplication of terms and speak of them from their first appearance as constituting the *pericardial region of the coelom.*

The relationships of the pericardial region of the coelom in embryos of 24 hours can most readily be grasped from a study of transverse sections. Figure 50, B, shows the great dilation of the coelom on either side of the anterior intestinal portal as compared with its condition farther caudad (Fig. 50, C). Where the splanchnic mesoderm lies closely applied to the entoderm at the lateral margins of the portal, it is noticeably thickened. It is from these areas of thickened splanchnic mesoderm that the paired primordia of the heart will later arise (Fig. 59).

In entire embryos of this age the thickened splanchnic mesoderm can be made out as a dark band lying close against the crescentic entodermal border of the anterior intestinal portal (Fig. 49). If the preparation is favorably stained the boundaries of the pericardial regions of the coelom can be traced (see Fig. 49). Following mesiad from the easily located thickened areas, the mesodermic borders can be seen to extend from either side, parallel to the entodermic margins of the portal, nearly to the midline. They then turn cephalad for a short distance. When they encounter the ectodermal fold which constitutes the posterior boundary of the subcephalic pocket, they swing laterad parallel with it and can be traced outside the embryonic region where they constitute the cephalic borders of the anterior horns of the mesoderm (see also Fig. 60, *A*).

The portion of the coelom, the borders of which we have just located between the subcephalic pocket and the anterior intestinal portal, is an important landmark from another standpoint than the part it is destined to play in the formation of the pericardial region. It is the most cephalic part of the coelom, for there is no coelom in the head. In the cephalic region, as we have seen, the mesoderm is not aggregated into definite masses or coherent cell layers, but derived from cells which migrate into the region. It will be recalled that these migrating cells are termed mesenchymal cells, or collectively mesenchyme. It should be borne in mind that the term is a general one, and later in development we shall encounter mesenchymal aggregations in many other regions of the body.

By careful focusing on the whole-mount the mesenchyme of the head can be seen as an indefinite mass lying between the superficial ectoderm and the entoderm of the foregut (Fig. 49). The distribution of the mesenchymal cells is easily seen in moderately magnified sections (Fig. 50, *A*). Their characteristic irregularity of shape, correlated with the fact they were carrying out active ameboid movements when they were alive, can be seen only when they are looked at under relatively high magnification (Fig. 62).

The Area Vasculosa   In a 24-hour chick the boundary between the area opaca and area pellucida has the same appearance and significance as in chicks of 18 to 20 hours. There is, however, a very marked difference between the proximal portion of the area opaca adjacent to the area pellucida and the more distal portions of the area opace. The proximal region is much darker and has a somewhat mottled appearance (Fig. 48). The greater density of this region is due to its invasion by mesoderm which makes it thicker and therefore more opaque in transmitted light (Fig. 50, *D*). The boundary between the inner and outer zones of the area opaca is established by the extent to which the mesoderm has grown peripherally. The distal zone is called the *area opaca vitellina* because the yolk alone underlies it. The proximal zone into which mesoderm has grown is known as the *area opaca vasculosa*, because it is from the mesoderm in this region that the yolk-sac blood vessels arise. The mottled appearance of this region is due to the aggregation of mesoderm into cell clusters, or *blood islands*, which mark the initial step in the formation of blood vessels and blood corpuscles. Later in development the formation of blood islands and vessels extends in toward the body of the embryo from its place of earliest appearance in the area opaca and involves the mesoderm of the area pellucida. The histological nature of the blood islands will be taken up in connection with later stages where their development is more advanced.

CHAPTER   7

CHANGES BETWEEN TWENTY-FOUR AND THIRTY-THREE HOURS OF INCUBATION

In dealing with developmental processes the choosing of stages for detailed consideration is more or less arbitrary and largely determined by the phenomena one seeks to emphasize. There is no stage of development which does not show something of interest but it is impossible in brief compass to take up at length more than a few selected ones. Nevertheless it is important not to lose the continuity of the processes involved. By calling attention to some of the more important intervening changes, this brief chapter aims to bridge the gap between the 24-hour stage and the 33-hour stage of the chick, both of which are taken up in some detail.

Closure of the Neural Groove   In comparison with 24-hour chicks, entire embryos of 27 to 28 hours of incubation (Fig. 52) show marked advances

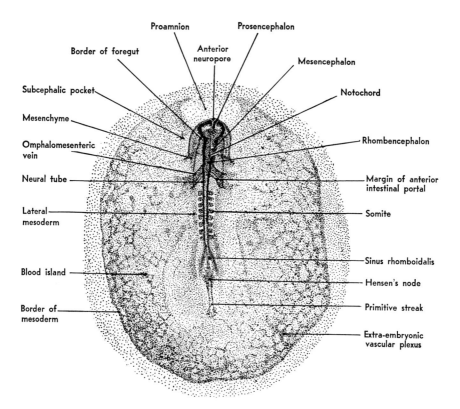

Proamnion

Prosencephalon

Border of foregut

Anterior
neuropore

Mesencephalon

Subcephalic pocket

Notochord

Mesenchyme

Omphalomesenteric
vein

Rhombencephalon

Neural tube

Margin of anterior
intestinal portal

Lateral
mesoderm

Somite

Sinus rhomboidalis

Blood island

Hensen's node

Border of
mesoderm

Primitive streak

Extra-embryonic
vascular plexus

Fig. 52. Dorsal view ($\times$14) of entire chick embryo having eight pairs of somites (about 27 to 28 hours' incubation).

in the development of the cephalic region. The head has elongated rapidly and now projects free from the blastoderm for a considerable distance, with a corresponding increase in the depth of the subcephalic pocket and in the length of the foregut.

In 24-hour chicks the anterior part of the neural plate is already folded to form the neural groove. Although the neural folds are at that stage beginning to converge middorsally, the groove nevertheless remains open throughout its length (Fig. 50, *A–C*). By 27 hours the neural folds in the cephalic region meet in the middorsal line, and their edges become fused.

The fusion which occurs is really a double one. Careful following of Fig. 59, *A–E*, will aid greatly in understanding the process. Each neural fold consists of two components, one of which is thickened neural plate ectoderm and the other of which is unmodified superficial ectoderm (Fig. 59, *A*). When the neural folds meet in the middorsal line (Fig. 59, *B, C*), the mesial, neural plate components of the two folds fuse with each other, and the outer layers of unmodified ectoderm also become fused (Fig. 59, *D*). Thus in the same process the neural groove becomes closed to form the neural tube, and the superficial ectoderm closes over the place formerly occupied by the open neural groove. Shortly after this double fusion the neural tube and the superficial ectoderm become somewhat separated from each other leaving no hint of their former continuity (Fig. 59, *E*).

Differentiation of the Brain Region   By 27 hours of incubation the cephalic part of the neural tube is markedly enlarged as compared with the more caudal parts. Its thickened walls and dilated lumen mark the region which will develop into the brain. The undilated posterior part of the neural tube gives rise to the spinal cord. Three divisions, the *primary brain vesicles*, can be distinguished in the enlarged cephalic region of the neural tube (Fig. 52). Occupying most of the rostral part of the head is a conspicuous dilation known from its position as the *forebrain* or *prosencaphalon*. Posterior to the prosencephalon and marked off from it by a constriction is the *midbrain* or *mesencaphalon*. Posterior to the mesencephalon with only a very slight constriction marking the boundary is the *hindbrain* or *rhombencephalon*. The rhombencephalon is continuous posteriorly with the cord region of the neural tube without any definite point of transition.

In somewhat older embryos (Fig. 53) the lateral walls of the prosencephalon become outpocketed to form a pair of rounded dilations known as the *primary optic vesicles*. When the optic vesicles are first formed there is no constriction between them and the lateral walls of the prosencephalon, and the lumen of each optic vesicle communicates mesially with the lumen of the prosencephalon without any definite line of demarcation.

The relation of the notochord to the divisions of the brain is of importance in later developmental processes. The notochord extends anteriorly as far as a depression in the floor of the prosencephalon known as the *infundibulum* (Fig. 53). Therefore, the rhombencephalon, mesencephalon, and that part of the prosencephalon posterior to the infundibulum lie immediately dorsal to the notochord (are epichordal) while the parts of the prosencephalon cephalic to it project rostral to the notochord (are prechordal).

The Anterior Neuropore   The closure of the neural folds takes place first near the anterior end of the neural groove and progresses thence both caphalad and caudad. At the extreme rostral end of the brain closure is delayed. As a result the prosencephalon remains for some time in communication with the outside through an opening called the *anterior neuropore*. The anterior neuropore is still open in chicks of 27 hours (Fig. 52). In embryos of 30 hours the neuropore appears much narrowed (Fig. 53), and by 33 hours it is almost closed (Fig. 55). A little later it becomes entirely closed but leaves for some time a scar-like fissure in the anterior wall of the prosencephalon. The anterior neuropore does not give rise to any definite brain structure. It is important simply as a landmark in brain topography. Long after it has disappeared as a definite opening, the scar left by its closure serves to mark the point originally most anterior in the developing brain.

The Sinus Rhombiodalis   The brain merges caudally without any definite line of demarcation into the region of the neural tube destined to become the spinal cord. After 27 hours of incubation the neural tube as far caudally as somite formation has progressed is completely closed and of nearly uniform diameter. Caudal to the most posterior somites the neural groove is still open, and the neural folds diverge to either side of Hensen's

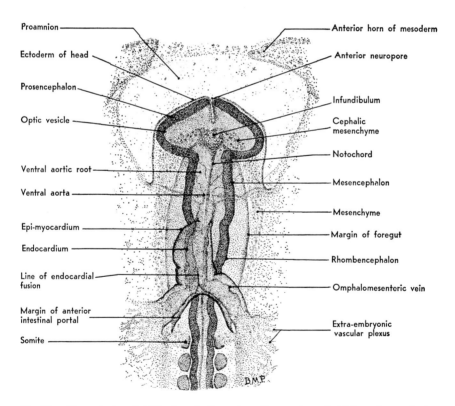

Fig. 53.   Ventral view (×47) of cephalic and cardiac region of chick embryo of nine somites (about 29 to 30 hours' incubation).

node (Fig. 52). In their later growth caudad the neural folds converge toward the midline and form the lateral boundaries of an unclosed region at the posterior extremity of the neutral tube known because of its shape as the *sinus rhomboidalis* (Fig. 55). Hensen's node and the primitive pit lie in the floor of this as yet unclosed region of the neural groove and subsequently are enclosed within it when the neural folds here finally fuse to complete the neural tube.

This process in the chick is homologous with the enclosure of the blastopore by the neural folds in lower vertebrates (Fig. 24, *E*). In forms where the blastopore does not become closed until after it is surrounded by the neural folds, it for a time constitutes an opening from the neural canal into the primitive gut known as the *neurenteric canal*. In the chick the fact that there is never an open blastopore precludes the establishment of an open neurenteric canal, but the primitive pit represents its homolog.

Formation of Additional Somites   The division of the dorsal mesoderm to form somites begins to be apparent in embryos of about 22 hours (Fig. 45). By the end of the first day three or four pairs of somites have been cut off (Fig. 48). As development progresses new somites are added caudal to those first formed. In embryos which have been incubated about 27 hours, eight pairs of somites have been established (Fig. 52).

Attention has already been called to the fact that because of such factors as overnight retention of eggs before laying, or different amounts of incubation before they were removed from the nest, there is great variability in the developmental progress within a group of eggs incubated for a given time in the laboratory. To meet this situation Hamburger and Hamilton (1951) have worked out a series of numbered stages of developmental progress for chick embryos handled under carefully controlled conditions. These stages are of great value in research because they facilitate more critical comparisons of the results of studies carried out by different investigators under varying laboratory conditions. For the beginner such elaborate staging is unnecessary and hard to remember. A simple alternative is the number of pairs of somites that have been formed. Chicks with a given number of pairs of somites will be found to be very much alike in developmental progress. Somite counts are easy to make and serve as a satisfactory basis of comparison. The "incubation age" is frankly merely a convenient approximation.

It was formerly believed that some new somites were formed anterior to the first pair. The experiments of Patterson indicate quite definitely that the first pair of completely formed somites remains the most anterior and that all the new somites are added posterior to them. The experiments referred to were carried out on eggs which had been incubated up to the time of the formation of the first somite. With thorough aseptic precautions the eggs were opened and the first somite marked, in some cases by injury with an "electric needle," in other cases by the insertion of a minute glass pin. Following the operation the shell was closed by sealing over the opening a piece of egg shell of appropriate size. After being incubated again for varying lengths of time the eggs were reopened. In all cases the injured first somite was still the most anterior complete somite. All the new somites except the incomplete "head somite" had appeared caudal to the first pair of somites formed.

Lengthening of the Foregut   Comparison of the relations of the crescentic margin of the anterior intestinal portal in embryos between 24 and 30 hours shows it occupying progressively more caudal positions. (Note its location in relation to the first somites in Figs. 60 and 61.) This change in the position of the anterior intestinal portal is the result of two distinct growth processes. The margins of either side of the portal are constantly converging toward the midline where they become merged. This lengthens the foregut by adding to its floor and thereby displacing the crescentic margin of the portal caudad. At the same time the structures cephalic to the anterior intestinal portal are elongating rapidly so that the portal becomes more and more remote from the anterior end of the embryo. This results in further lengthening of the foregut.

As a result of these two processes the space between the subcephalic pocket and the margin of the anterior intestinal portal is also elongated (Fig. 60). This is of importance in connection with the formation of the heart; for it is into this enlarging space that the pericardinal portions of the coelom extend and within it that the heart comes to lie.

Appearance of the Heart and Omphalomesenteric Veins   Although the early steps in the formation of the heart take place in embryos of this range, detailed consideration of them has been deferred to be taken up in connection with later stages when conditions in the circulatory system as a whole are more advanced. Nevertheless the location of the cardiac primordia in the embryonic body should be noted at this time.

In dorsal views of entire embryos the heart is largely concealed by the overlying rhombencephalon (Fig. 52), but it may readily be made out by viewing the embryo from the ventral surface (Fig. 53). At this stage the heart is a nearly straight tubular structure lying in the midline ventral to the foregut. Its midregion has noticeably thickened walls and is somewhat dilated. Anteriorly the heart is continuous with a large medial vessel, the *ventral aorta;* posteriorly it is continuous with the paired *omphalomesenteric veins.* The fork formed by the union of the omphalomesenteric veins in the posterior part of the heart lies immediately cephalic to the crescentic margin of the anterior intestinal portal, the veins lying within the fold of entoderm which constitutes its margin.

Organization in the Area Vasculosa   The extra-embryonic vascular area at this stage is undergoing rapid enlargement and presents a netted appearance instead of being mottled as in the earlier embryos. The peripheral boundary of the area vasculosa is definitely marked by a dark band, the precursor of the *sinus terminalis (marginal sinus).* The netted appearance of the area vasculosa is due to the extension and anastomosing of blood islands. The formation of the network is a step in the organization of a plexus of blood vessels on the yolk surface which will later be the means of absorbing and transferring food material to the embryo. The afferent yolk-sac or vitelline circulation is established in the next few hours of incubation when this plexus of vessels developing on the yolk surface comes into communication with the omphalomesenteric veins already developing within the embryo and extending laterad. The efferent vitelline circulation is established somewhat later when the omphalomesenteric arteries arise from the aorta of the embryo and become connected with the yolk-sac plexus (cf. Figs. 52, 55, and 67).

CHAPTER 8

STRUCTURE OF CHICKS BETWEEN THIRTY-THREE AND THIRTY-EIGHT HOURS OF INCUBATION

Chicks which have been incubated from 33 to 38 hours are in a favorable stage to show some of the fundamental steps in the formation of the central nervous system, and of the circulatory system. In this chapter, therefore, attention has been concentrated on these two systems.

During this period of incubation there are also changes in the foregut region and in the somites, and differentiation in the intermediate mesoderm which presages the formation of the urinary organs. Consideration of these structures has, however, been deferred until their development has progressed somewhat farther.

Divisions of the Brain and Their Neuromeric Structure   The metameric arrangement of structures, which is so striking a feature in the body organization of all vertebrates, is masked in the head region of the adult by super-imposed specializations. In the brain of young vertebrate embryos, however, the metamerism is still indicated. Dissections of the neural plate of chicks at the end of the first day of incubation shows a series of eleven enlargements marked off from each other by constrictions (Fig. 54, *A*). Concerning the precise homologies of individual enlargements with specific neuromeres in other forms there is not complete agreement. The controversies center about the question of neuromeric fusions in the more rostral parts of the brain. For the beginning student the fact that metamerism is present is to be em-phasized rather than the controversies concerning the homologies of neuro-meres. With the reservation that some of the more rostral enlargements may represent fusions of more than one neuromere, the series of enlargements seen in the brain region of the chick may be regarded as neuromeric. For convenience in designation the neuromeres are numbered beginning at the rostral end.

With the closure of the neural tube and the establishment of the three primary brain vesicles we can begin to trace the fate of the various neuro-meric enlargements in the formation of the brain regions. The three most rostral neuromeres form the *prosencephalon;* neuromeres four and five are incorporated in the *mesencephalon;* and neuromeres six to eleven in the *rhombencephalon* (Fig. 54, *B*). Anteriorly the interneuromeric constrictions soon disappear except for two; namely, the one between the prosencephalon and mesencephalon and the one between the mesencephalon and rhomb-encephalon. The rhombencephalic neuromeres, however, remain clearly marked for a considerable period.

By about 33 hours of incubation the optic vesicles are established as paired lateral outgrowths of the prosencephalon. They soon extend to occupy the full width of the head (Fig. 54, *C* and Fig. 55). The distal portion of each of the vesicles thus comes to lie closely approximated to the super-ficial ectoderm, a relationship of importance in their later development. At first the cavities of the optic vesicles (opticoeles) are broadly confluent with the cavity of the prosencephalon (prosocoele). Somewhat later constrictions appear which mark more definitely the boundaries between the optic vesicles and the prosencephalon (Fig. 54, *D* and Fig. 56).

Concurrently there has been formed in the floor of the prosencephalon a depression, known because of its perculiar shape as the *infundibulum* (Figs. 57 and 64). The infundibular region is the site of important changes later in development. At this stage, conditions are not sufficiently advanced to warrant more than calling attention to its origin as a depression in the

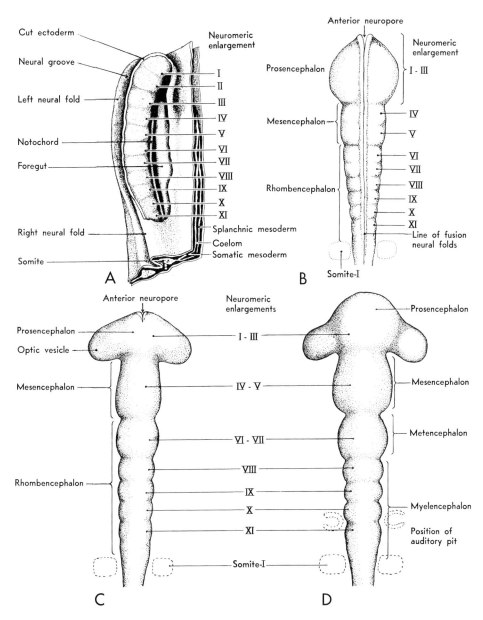

Fig. 54. Diagrams to show the neuromeric enlargements in the brain region of the neural tube. (*A*) Lateral view of neural plate from dissection of chick of 4 somites (24 hours). (*B*) Dorsal view of brain dissected out of 7-somite (26- to 27-hour) embryo. (*C*) Dorsal view of brain from 10-somite (30-hour) embryo. (*D*) Dorsal view of brain from 14-somite (36-hour) embryo. (*Based on figures by Hill.*)

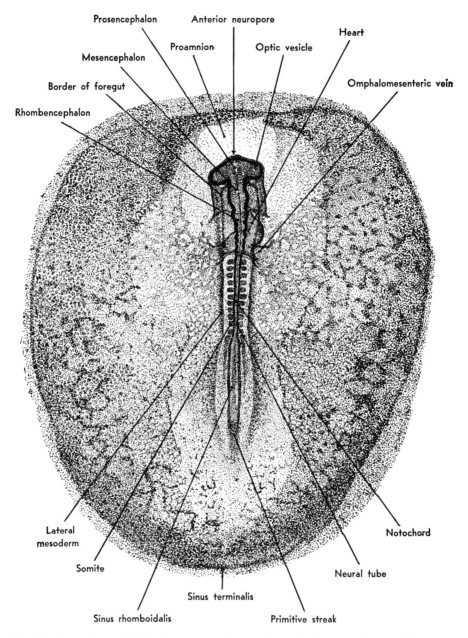

Fig. 55.   Dorsal view (×17) of an entire chick embryo of 12 somites (about 33 hours' incubation).

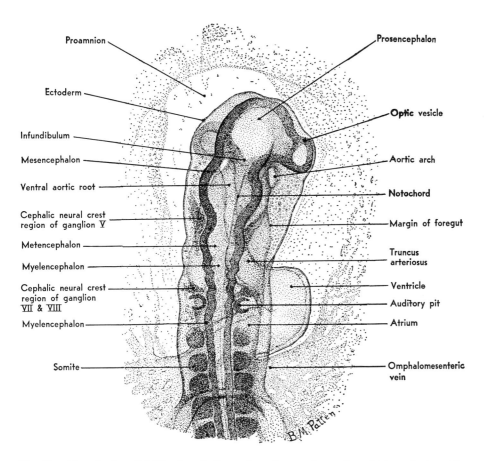

Proamnion — Prosencephalon

Ectoderm —

Optic vesicle

Infundibulum —

Mesencephalon — Aortic arch

Ventral aortic root — Notochord

Cephalic neural crest region of ganglion V — Margin of foregut

Metencephalon — Truncus arteriosus

Myelencephalon —

Cephalic neural crest region of ganglion VII & VIII — Ventricle

Auditory pit

Myelencephalon — Atrium

Somite — Omphalomesenteric vein

Fig. 56. Dorsal view (×40) of cephalic and cardiac regions of a chick embryo with the 17th somite just forming (about 38 hours' incubation).

floor of the prosencephalon and its positional relation to the anterior end of the notochord (Figs. 53 and 56).

In chicks of about 38 hours indications of the impending division of the three primary vesicles to form the five regions characteristic of the adult brain are already beginning to appear. In the establishment of the five-vesicle condition of the brain, the prosencephalon is subdivided to form the *telencephalon* and *diencephalon,* the mesencephalon remains undivided, and the rhombencephalon divides to form the *metencephalon* and *myelencephalon.*

The division of the prosencephalon into telencephalon and diencephalon is not well marked until a much later stage of development, but the median enlargement at this stage extending rostrad beyond the level of the optic vesicles indicates where the telencephalon will be established (Fig. 54,

D). The optic vesicles and that part of the prosencephalon lying between them go into the diencephalon.

The mesencephalon, as stated above, undergoes no subdivision, the original mesencephalic region of the three-vesicle brain giving rise to the mesencephalon of the adult. This region of the brain does not undergo any marked differentiation until relatively late in development.

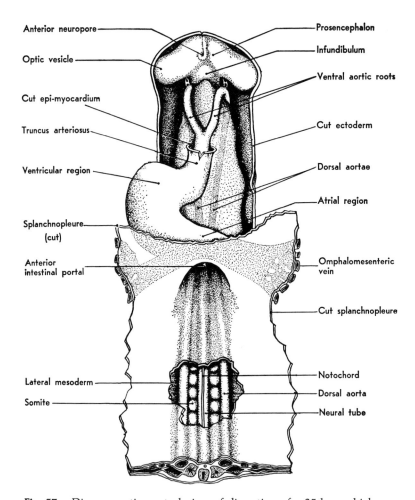

Fig. 57.   Diagrammatic ventral view of dissection of a 35-hour chick embryo. The splanchnopleure of the yolk-sac cephalic to the anterior intestinal portal, the ectoderm of the ventral surface of the head, and the mesoderm of the pericardial region have been removed to show the underlying structures. Figure 64 should be referred to for the relations of the pericardial mesoderm. (*Modified from Prentiss*)

At this stage the beginning of the subdivision of the rhombencephalon is clearly indicated (Fig. 54, *D* and Fig. 56). The two most anterior neuromeres of the original rhombencephalon form the metencephalon, and the posterior four neuromeres are incorporated in the myelencephalon.

The Auditory Pits.   As is the case with the central nervous system, the organs of special sense arise early in development. The appearance of the optic vesicles which later become the sensory parts of the eyes has already been noted. The first indication of the formation of the sensory part of the ear becomes evident at about 35 hours of incubation. At this age a pair of thickenings termed the *auditory placodes* arise in the superficial ectoderm of the head. They are situated on the dorso-lateral surface opposite the most posterior inter-neuromeric constriction of the myelencephalon. By 38 hours of incubation (Fig. 56) the auditory placodes have become depressed below the general level of the ectoderm and form the walls of a pair of cavities, the auditory pits. When first formed, the walls of the auditory pits are directly continuous with the superficial ectoderm and their cavities are widely open to the outside. In later stages the openings into the pits become narrowed and finally closed, so that the pits become vesicles lying between the superficial ectoderm and the myelencephalon. As yet they have no connection with the central nervous system.

Formation of Extra-embryonic Blood Vessels   In dealing with the circulation of the chick we must recognize at the outset two distinct circulatory arcs of which the heart is the common center. One complete circulatory arc is established entirely within the body of the embryo. A second arc is established which has a rich plexus of terminal vessels located in the extra-embryonic membranes enveloping the yolk. These are the vitelline vessels. The vitelline vessels communicate with the heart over main vessels which traverse the embryonic body. The chief distribution of the vitelline circulation is, however, extra-embryonic. Later in development there arises a third circulatory arc involving another set of extra-embryonic vessels in the allantois, but with that we have no concern until we take up later stages. Neither the intra-embryonic nor the vitelline circulatory channels have been completed in the younger embryos of this age range, but the heart and many of the main vessels have made their appearance.

The formation of extra-embryonic blood vessels is presaged by the appearance of blood islands in the vascular area of chicks toward the end of the first day of incubation. (Review Figs. 48 and 50, *D*.) Figure 58 shows the differentiation of blood islands to form primitive blood corpuscles and blood vessels. At the time of their first appearance the blood islands are irregular clusters of mesoderm cells lying in intimate contact with the yolk-sac entoderm (Fig. 58, *A*). When the lateral mesoderm becomes split, forming the somatic and splanchnic layers with the coelom between, the blood islands lie in the splanchnic mesoderm adjacent to the entoderm. In embryos of 3 to 5 somites, fluid-filled spaces begin to appear in the blood islands with the result that in each blood island the peripheral cells are separated from the

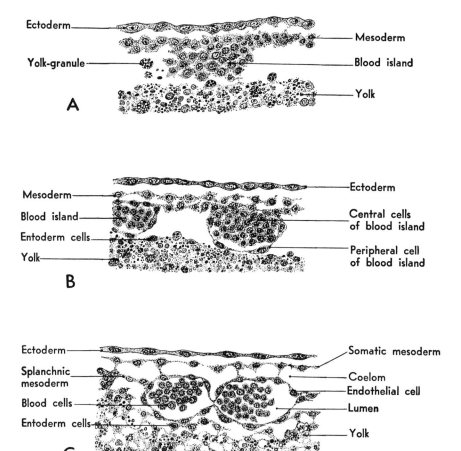

Fig. 58. Drawings to show the cellular organization of blood islands at three stages in their differentiation. By referring to Fig. 50, *D*, the small areas here represented can be located in relation to the structure of the embryo as a whole. (*A*) From blastoderm of 18-hour chick. (*B*) From blastoderm of 24-hour chick. (*C*) From blastoderm of 33-hour chick.

central ones (Fig. 58, *B*). As the fluid accumulates and the spaces expand, the peripheral cells become flattened and pushed outward, but they remain adherent to each other and completely enclose the central cells. At this stage the single layer of peripheral cells may be regarded as constituting the endothelial wall of a primitive blood channel (Fig. 58, *C*). Extension and anastomosis of neighboring blood islands which have undergone similar differentiation result in the establishment of a network of communicating vessels. Meanwhile the cells enclosed in the primitive blood channels have become separated from each other and rounded. They soon come to contain hemoglobin and constitute the primitive blood corpuscles. The fluid accumulated in the blood islands serves as a vehicle in which the corpuscles are suspended and will later be conveyed along the vessels.

The differentiation of the blood islands in the manner described begins first in the peripheral part of the area vasculosa and from there extends toward the body of the embryo. By 33 to 35 hours of incubation the extra-embryonic vascular plexus has extended inward and made connection with the omphalomesenteric veins which, originating within the body of the embryo, have grown outward (Fig. 55). Thus are established the afferent vitelline channels.

The efferent vitelline channels have not yet appeared, and there is no circulation of the blood corpuscles which are being formed in the area vasculosa. The developing intra-embryonic blood vessels contain fluid but no corpuscles until the extra-embryonic circuit is completed. The embryo meanwhile draws its nutrition from the yolk by direct absorption.

Formation of the Heart   The structural relations of the heart and the way in which it is derived from the mesoderm can be grasped only by the careful study of sections through the cardiac region in several stages of development (Fig. 59). The fact that the heart, itself an unpaired structure, arises from paired primordia, which at first lie widely separated on either side of the midline, is likely to be troublesome unless its significance is understood at the outset. The paired condition of the heart at the time of its origin is due to the fact that the early embryo lies open ventrally, spread out on the yolk surface. The primordia of all ventral structures which appear at an early age are thus at first separated, and lie on either side of the midline. As the embryo develops, a series of foldings undercut it and separate it from the yolk. This folding-off process at the same time establishes the ventral wall of the gut and the ventral body-wall of the embryo by bringing together in the midline the structures formerly spread out to right and left. The primordia of the heart arise in connection with layers which are destined to form ventral

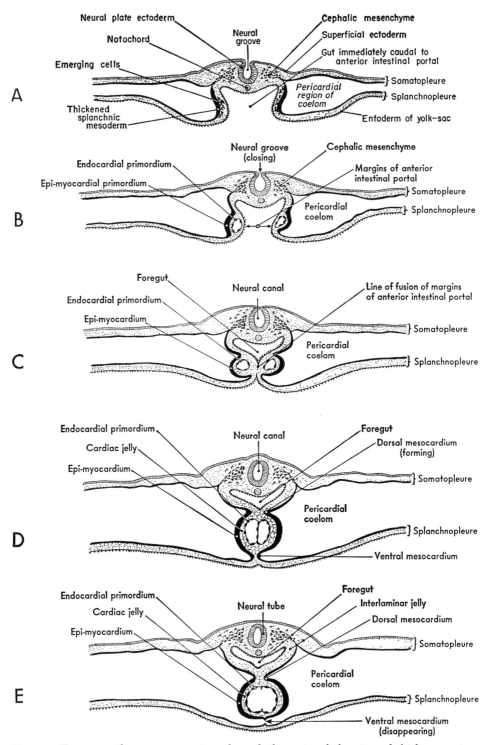

Fig. 59. Diagrams of transverse sections through the pericardial region of chicks at various stages to show the formation of the heart. For location of the sections consult Fig. 60. (*A*) At 25 hours; (*B*) at 26 hours; (*C*) at 27 hours; (*D*) at 28 hours; (*E*) at 29 hours.

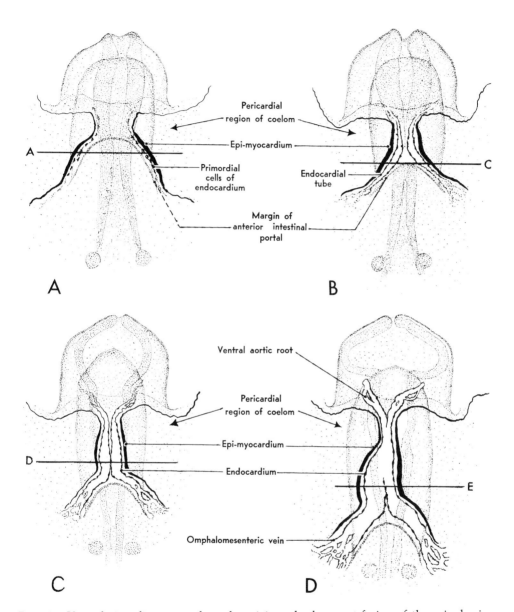

Fig. 60.   Ventral-view diagram to show the origin and subsequent fusion of the paired primordia of the heart. The lines *A, C, D,* and *E* indicate the planes of the sections diagrammed in Fig. 59, *A, C, D,* and *E,* respectively. (*A*) Chick of 25 hours; (*B*) chick of 27 hours; (*C*) chick of 28 hours; (*D*) chick of 29 hours.

parts of the embryo, but at a time when these layers are still spread out on the yolk. As the embryo is completed ventrally, the paired primordia of the heart are brought together in the midline and become fused (Figs. 59 and 60).

Although, as we have seen, prospective cardiac tissue can be recognized much earlier, the first definite structural indication of heart formation appears in transverse sections passing through a 25-hour chick immediately caudal to the anterior intestinal portal. Where the splanchnopleure of either side bends toward the midline along the lateral margin of the intestinal portal, there is a marked regional thickening in the splanchnic mesoderm (Figs. 49, 59, A, and 60, A). This pair of thickenings indicates where there has been rapid cell proliferation preliminary to the differentiation of the heart. Loosely associated cells can already be seen somewhat detached from the mesial face of the mesodermal layer. These cells soon become organized to form the *endocardial primordia*.

In a chick of about 26 hours, sections through a corresponding region show distinct differentiation of the endocardial and epimyocardial primordia (Fig. 59, B). The endocardial primordia are a pair of delicate tubular structures, a single cell in thickness, lying between the entoderm and mesoderm. They arise from the cells seen separating from the adjacent thickened mesoderm in the 25-hour chick. As their name indicates they are destined to give rise to the endothelial lining of the heart. By far the greater part of each of the original mesodermic thickenings becomes applied to the lateral aspects of the endocardial tubes as the *epi-myocardial primordium* which is destined to give rise to the external coat of the heart (epicardium) and to the heavy muscular layers of the heart (myocardium).

In chicks of 27 hours the lateral margins of the anterior intestinal portal have been undergoing concrescence, thus lengthening the foregut caudally and at the same time elongating the pericardial region. In this process the lateral margins of the portal swing in to meet each other and merge in the midline, and the endocardial tubes of the right and left side are brought toward each other beneath the newly completed floor of the foregut (Figs. 59, C, and 60, B). In the 28-hour chick the endocardial primordia are approximated to each other (Figs. 59, D, and 60, C), and by 29 hours they fuse in their midregion to form a single tube (Figs. 59, E and 60, D).

At the same time the epi-myocardial areas of the mesoderm are brought together first ventrally (Fig. 59, D) and then dorsally to the endocardium (Fig. 59, E). Where the splanchnic mesodermal layers of the opposite sides of the body become apposed to each other dorsal and ventral to the heart, they form double-layered supporting membranes called, respectively, the *dorsal mesocardium* and the *ventral mesocardium*. The ventral mesocardium is a transitory structure, disappearing almost as soon as it is formed (Fig. 59, *E*). The dorsal mesocardium, although the greater part of it disappears in the next few hours of incubation, persists in embryos of the early part of the age range under consideration, suspending the heart in the pericardial region of the coelom. The general relations in the cardiac region at 33 hours of incubation are shown in section in Fig. 63, *C*. The heart by this stage is enlarged and displaced somewhat to the right of the midline, but its fundamental relations are otherwise the same as in a 29-hour embryo (cf. Figs. 59, *E*, and 63, *C*).

Under high magnification, careful study of well-stained sections of the developing heart at this stage reveals the presence of *cardiac jelly* between the epi-myocardium and the endocardium. In living embryos this cardiac jelly appears homogeneous and structureless, but, nevertheless, its presence can be inferred from the way it holds the endocardium and myocardium together and makes them move synchronously when the heart begins to pulsate. In fixed and stained material the cardiac jelly appears as a very delicate, noncellular coagulum. It is of interest functionally not only because of the way it gives mechanical unity to the endocardium and the myocardium during early stages, but also because it serves later as the substratum on which migrating cells move in and form a well-organized tissue, knitting together these two primordial layers of the heart. A study of Fig. 59, *E*, will make it evident that the cardiac jelly is directly continuous with the interlaminar jelly with which we have already become acquainted as the substratum in which the migrating mesenchymal cells are supported. Davis (1924) realized the importance of this substance in the heart long before the existence of a similar substance with a like significance was recognized in the embryonic body as a whole; hence the special term cardiac jelly antedated the more general term interlaminar jelly.

The gross shape of the heart and its positional relations to other structures can be seen readily in entire embryos. The fusion of the paired cardiac primordia establishes the heart as a nearly straight tubular structure. It lies at the level of the rhombencephalon in the midline, ventral to the foregut (Fig. 53). By 33 hours of incubation the midregion of the heart is considerably dilated and bent to the right (Fig. 55). At 38 hours the heart is bent so far to the right that it extends beyond the lateral body margin of the embryo (Fig. 56). This bending process is correlated with the rupture of the dorsal mesocardium at the midregion of the heart. The breaking through of the dorsal and ventral mesocardia is of interest aside from the fact that it leaves

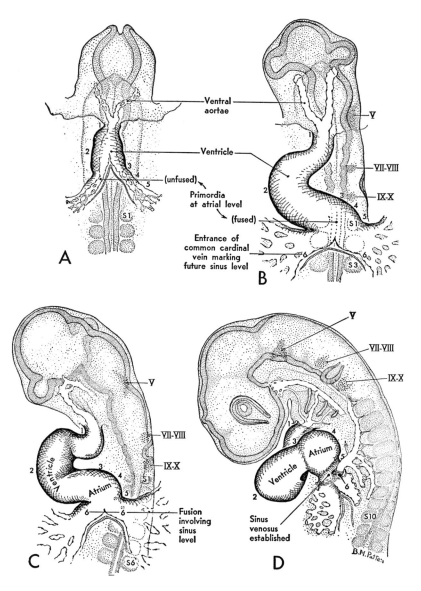

Fig. 61. The formation of the fundamental regions of the chick heart by pro-gressive fusion of its paired primordia. (*A*) Ventral view at the 9-somite stage, when the first contractions appear. The ventricular part of the heart is the only region where the fusion of the paired primordia has occurred and the myocar-dial investment has been formed. (*B*) Ventral view at the 16-somite stage, when the blood first begins to circulate. The atrium and ventricle have been estab-lished, but the sinus venosus exists only as undifferentiated primordial channels, still paired and still lacking myocardial investment. (*C*) Ventrosinistral view at

the heart free to undergo changes in shape. It makes the right and left coelomic chambers confluent, the pericardial region thus being the first part of the coelom to acquire the unpaired condition characteristic of the adult.

Although there are as yet no sharply bounded subdivisions of the heart, its fundamental regions are beginning to be shaped. Taking them in order from what is going to be the intake end of the heart toward its discharging end, they are: sinus venosus, atrium, ventricle, and truncus arteriosus. At the 36- to 38-hour stage the *sinus venosus* is only suggested. It is represented by the still paired primordia where the common cardinals enter the omphalomesenteric veins, and they in turn are becoming confluent with each other to enter the atrial portion of the tubular heart (Figs. 61, *C*, and 64). The *atrium* is held close beneath the posterior part of the foregut by a persisting portion of the dorsal mesocardium. The *ventricle* is the part of the cardiac tube that makes a U-shaped bend to the right, which brings it into clear view at the side of the body in whole-mounts (Fig. 56). Swinging back to the midline the ventricle is narrowed to form the discharging part of the heart known as the *truncus arteriosus* (Fig. 57). The part of the ventricle that narrows abruptly to give rise to the truncus arteriosus is spoken of as the *ventricular cone* or more briefly as the *conus*. This, rather than being a basic cardiac region, is really just a zone of transition for which it is convenient to have a name.

From the way the paired cardiac primordia are at first located on either side of the anterior intestinal portal it is evident that they can fuse with each other only as the "flooring-in" of the foregut progresses (cf. Fig. 59, *B–D*). That means that the formation of the basic regions of the young tubular heart is a sequential process. The truncoventricular part of the heart is formed first (Fig. 61, *A*). Then the atrium is added caudal to the ventricle (Fig. 61, *B*, *C*). Finally, in stages older than those here under discussion, the sinus venosus is added caudal to the atrium (Fig. 61, *D*). When we come to consider the way the heart starts to beat, we shall find that the sequential formation of its basic regions is most interestingly correlated with its characteristic physiology.

the 19-somite stage. Fusion of the paired primorida is just beginning to involve the sinus region. (*D*) Sinistral view at the 26-somite stage. The sinus venosus is definitely established and its investment with myocardium well advanced. To facilitate following the progress of fusion, Arabic numerals have been placed against approximately corresponding locations. The 6 is located at the point of entrance of the common cardinal vein as determined from injected specimens. [*From Patten and Kramer, Am. J. Anat.*, **53** (1933).]

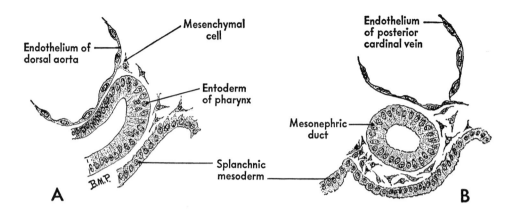

Fig. 62. High-power drawings contrasting the histology of vascular endothelium with other epithelial layers in the embryo. (*A*) From the cephalic region of a 36-hour chick. (*B*) From the mesonephric region of a 60-hour chick.

Formation of Intra-embryonic Blood Vessels   Concurrently with the establishment of the heart, blood vessels have been arising within the body of the embryo. Concerning the exact nature of the process of blood vessel formation there has been much disagreement. The weight of evidence now indicates that the early vessels are formed from mesodermal cells which lie in the path of their development. They grow by organization of cells *in situ* as a drain might be built from bricks already deposited along its projected course. In later stages vessels extend by the formation of bud-like outgrowths from their walls, as well as by organization of cells in their path of development. When first formed, the blood vessel walls are but a single cell in thickness. There is no structural differentiation between arteries and veins until a considerably later period. Recognition of specific vessels depends wholly, therefore, on determining their course and relationships. It is, however, easy to distinguish blood vessels from other tubular structures. Blood vessels are always lined by a single layer of thin, flattened cells called *endothelium*. Other hollow organs in young embryos, as for example the gut or the mesonephric duct, will be found lined by thicker epithelial layers made up of cuboidal or columnar cells (Fig. 62).

The large vessels connecting with the heart are the first of the intra-embryonic channels established. At the cephalic end of the pericardial cavity the epi-myocardial covering of the truncus arteriosus is reflected to become continuous with the general mesodermal lining of the pericardial coelom (Fig. 64). The endothelial lining of the truncus is continued cephalad beneath the foregut as the *ventral aorta*. Almost immediately the ventral aorta bifurcates to form the paired *ventral aortic roots* (Fig. 57). At the cephalic end

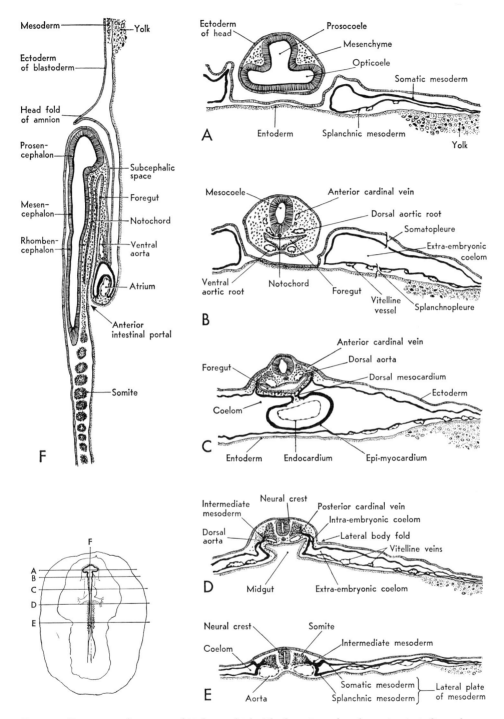

Fig. 63. Diagrams of sections of 33-hour chick. The location of each section is indicated on a small outline sketch of the entire embryo.

of the foregut the ventral aortic roots turn dorsad, curve around the gut, and then extend caudad as the paired *dorsal aortae* (Figs. 63, *B*, and 64). Few conspicuous branches arise from the aortae at this stage, but as development progresses, branches extend to the various parts of the embryo and the aortae become the main efferent conducting vessels of the embryonic circulation.

The curved vessels connecting the ventral aortic roots with the dorsal aortic roots are the *aortic arches*. At this stage there is but a single pair, the first of the series of six which will appear as development progresses. Both the ventral aortic roots at the outlet end, and the omphalomesenteric veins at the intake end, are direct continuations of the paired endocardial primordia of the heart. The epi-myocardial coat is formed about the original endothelial tubes only where they are fused in the region destined to become the heart.

During early embryonic life the cardinal veins are the main afferent vessels of the intra-embryonic circulation. The cardinal trunks are paired vessels, symmetrically placed on either side of the midline. There are two pairs, the *anterior cardinal veins* which return the blood to the heart from the cephalic region of the embryo, and the *posterior cardinal veins* which return the blood from the caudal region. The anterior and posterior cardinal veins of the same side of the body become confluent dorsal to the level of the heart, the vessels formed by their junction being the *common cardinal veins*. The right and left common cardinal veins turn ventrad, one on either side of the foregut, and enter the sino-atrial end of the heart along with the right and left omphalomesenteric veins, respectively (Fig. 64).

In chicks of 33 hours of incubation the anterior cardinal veins can usually be made out in sections (Fig. 63, *B, C*). By 38 hours the anterior and common cardinals are readily recognized. The posterior cardinals appear somewhat later than the anterior cardinals but are ordinarily discernible just caudal to the common cardinal veins by 33 to 35 hours and well established by 38 hours. For the sake of simplicity and clearness the cardinal veins have been represented in Fig. 64 somewhat larger and more regularly formed than they are in actual specimens. Like all the other blood vessels of the embryo, they arise as irregular anastomosing endothelial tubes, only gradually taking on the regularity of shape characteristic of fully formed vessels.

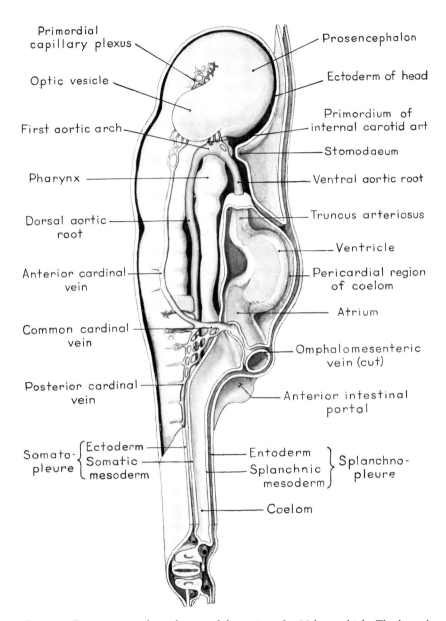

Fig. 64. Diagrammatic lateral view of dissection of a 38-hour chick. The lateral body-wall of the right side has been removed to show the internal structures. Note especially the relations of the pericardial region to that part of the coelom which lies farther caudally, and the small anastomosing channels of the developing posterior cardinal vein from which a single main vessel is later derived.

CHAPTER 9

CHANGES BETWEEN THIRTY-EIGHT AND FIFTY HOURS OF INCUBATION

Flexion and Torsion   Until 36 or 37 hours of incubation the longitudinal axis of the chick is straight except for slight fortuitous variations. Beginning at about 38 hours, processes are initiated which eventually change the entire configuration of the embryo and its positional relations to the yolk. These processes involve positional changes of two distinct types, flexion and torsion. As applied to an embryo, flexion means the bending of the body about a transverse axis, as one might bend the head forward at the neck, or the trunk forward at the hips. Torsion means the twisting of the body, as one might turn the head and shoulders in looking backwards without changing the position of the feet.

In chick embryos the first flexion of the originally straight body-axis takes place in the head region. Because of its location it is known as the cranial flexure. The axis of bending in the development of the cranial flexure is a transverse axis passing through the midbrain. The direction of the flexion is such that the forebrain becomes bent ventrally toward the yolk. Until the cranial flexure is well established, it is inconspicuous in dorsal views of whole-mounts, but even in its initial stages it appears plainly in lateral views.

To appreciate the correlation between the processes of flexion and torsion it is only necessary to bear in mind the relation of a chick of this age to the yolk. As long as the chick lies with its ventral surface closely applied to the yolk, the yolk constitutes a bar to flexion. Before extensive flexion can be carried out the chick must twist around on its side, i.e., undergo torsion, as a man lying face down turns on his side in order to flex his body.

Torsion begins in the cephalic region of the embryo and progresses caudad. The first indications of torsion appear almost as soon as the cranial flexure begins, and the two processes then progress synchronously. In the chick, torsion is normally carried out toward a definite side. The cephalic region of the embryo is twisted in such a manner that the left side comes to lie next to the yolk and the right side away from the yolk. The progress of torsion caudad is gradual, and the posterior part of the embryo remains prone on the yolk for a considerable time after torsion has been completed in the head region. Figure 56 shows the head of an embryo of about 38 hours in which the cranial flexure and torsion are just becoming evident. In chicks of about 43 hours (Fig. 65) the further progress of both flexion and torsion is well marked.

The processes of flexion and torsion thus initiated continue until the original orientation of the chick on the yolk is completely changed. As the body of the embryo becomes turned on its side, the yolk no longer impedes the progress of flexion. Following the accomplishment of torsion in the cephalic region, the cranial flexure becomes rapidly greater until the head is practically doubled on itself (Fig. 74). As development proceeds, torsion progresses caudad involving more and more of the body of the embryo. Finally the entire embryo comes to lie with its left side on the yolk. Concomitantly with the progress of torsion, flexion also appears farther caudally, affecting in turn the cervical, dorsal, and caudal regions. The series of flexions which accompany torsion bend the head and tail of the embryo toward each other so that its spinal axis becomes C-shaped (Fig. 83). The flexions which thus bend the embryo on itself are characteristic not of the chick alone, but of the embryos of reptiles, birds, and mammals generally. Flexion would seem to be correlated with the spatial limitations of the egg or the uterine cavity within which these embryos undergo their development. Certainly there is no appreciable flexion in embryos of fishes and amphibia which develop free in water with no imposed limitations of space. The tor-

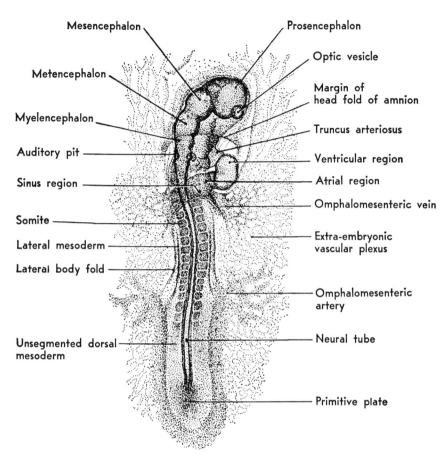

Fig. 65. Dorsal view (×20) of entire chick embryo having 19 pairs of somites (about 43 hours' incubation). Because of torsion the cephalic region appears in dextro-dorsal view.

sion which in the chick accompanies flexion is characteristic only of embryos developing on the surface of a very large yolk which would act as an impediment to flexion unless the embryo had first turned on its side.

Completion of the Vitelline Circulatory Channels   In chicks of 33 to 36 hours the omphalomesenteric veins have been established as postero-lateral extensions of the same endocardial tubes which are involved in the formation of the heart. As the omphalomesenteric veins are extending laterad, the vessels developing in the vitelline plexus are converging toward the embryo. Eventually the vitelline vessels attain communication with the omphalomesenteric veins establishing the return route from the vitelline vascular plexus to the heart.

The omphalomesenteric arteries which are destined to carry blood from the dorsal aortae to the vitelline plexus develop in embryos of about 40 hours (Fig. 65). Like the efferent vitelline channels, the afferent[1] vitelline channels have a dual origin. The proximal portions of the channels to the vitelline plexus arise within the embryo as branches of the dorsal aortae and extend peripherally. The distal portions of the channels arise in the extra-embryonic vascular area and extend toward the embryo. Obviously the vitelline circulation cannot begin until these two sets of channels become confluent. At first their connection with each other is by way of a network of small channels rather than by large vessels. These preliminary channels are formed as exceedingly small, freely anastomosing vessels extending from the aorta to communicate laterally with the extra-embryonic plexus. Later some of these channels become confluent, others disappear, and gradually definite main vessels, the omphalomesenteric arteries, are established. For some time after their formation, the omphalomesenteric arteries are likely to retain traces of the origin from a plexus of small channels and arise from the aorta by several roots (Fig. 75).

Beginning of the Circulation of Blood   We have traced the sequential formation of the various regions of the heart by the progressive fusion of its paired primordia, and studied the gradual development of the intra- and extra-embryonic blood vessels. It becomes a matter of particular interest, now, to see how this system goes into action. For without a functioning cardiovascular mechanism the embryo could not grow beyond the early stages of development when its mass is so small that it can depend on direct absorption of raw food material from the yolk and direct gaseous interchange through the porous shell.

---

[1] In dealing with main vascular channels near the heart the terms afferent and efferent are commonly used with reference to the heart as a pumping center. Thus an afferent channel is one conducting blood toward the heart, and an efferent channel is one conducting blood away from the heart. The same terms, however, may be quite properly employed with reference to other organs or structures. We speak, for example, of the afferent and efferent vessels of a renal glomerulus. It is quite possible, even, that the same vessels might be described in different terms according to the focus of attention at the moment. If, for example, we are considering the heart as the pumping center driving blood through the renal artery to the kidney, this blood channel may then be thought of as a branch of the main efferent circuit from the heart. If, however, we are studying the selective elimination of waste materials being carried out in the small terminal vessels within the kidney, we will describe blood flow in the renal artery as afferent with reference to these small renal vessels. In general when used with reference to organs other than the heart, a qualifying adjective will make the situation clear, as afferent *renal* vessel, or, as in this case, afferent *vitelline* channel.

As one might suspect, a mechanism as elaborate as the circulatory system does not go into full-scale operations all at once. Long before the actual circulation of blood commences, the heart has begun to pulsate tentatively and feebly. Its first contractions can usually be seen at the 9-somite stage (about 29 hours of incubation). At this age, it will be recalled that the fusion of the paired cardiac primordia has not progressed beyond the ventricular region (Fig. 61, A). These first beats are, therefore, in the ventricular myocardium. Their rhythmic recurrence is at first very slow and the amplitude of the contractions is small. The pulsations, watched in living embryos, are obviously far short of the power needed to set the blood in motion.

Within 3 or 4 hours of the time of the appearance of the first beats the rate and the amplitude of the pulsations are both strikingly increased. If one studies the structure of the heart at this stage, it is apparent that the fusion of the paired primordia has now established the atrium behind the ventricle (Fig. 61, B). By cutting across the heart of a living chick at this age it can be shown that the increased rate is due to the addition of the atrium behind the previously established ventricle. For in such experiments the atrial part of the heart continues to beat at the rate of the intact heart, whereas the ventricle, severed from the atrium, drops back to its initial very slow rate of pulsation (Fig. 66). This means that the myocardium in the newly formed atrium has a higher intrinsic contraction rate than that of the ventricle. When this more rapidly beating tissue is added behind the ventricle, it takes charge of the rate of the entire cardiac tube—becoming, as the physiologists say, its *pacemaker.*

Still later in development the sinus venosus is added behind the atrium (Fig. 61 C, D). As we shall see when we consider older embryos, similar cutting experiments carried out at these later stages show that the sinus myocardium has a contraction rate still higher than that of the atrium (Fig. 108). There is thus a cephalo-caudal gradient in the intrinsic rate of myocardial contraction. The importance of this situation is far-reaching. It means that the pacemaking center of the heart is always at its intake end. Therefore, when a wave of contraction starts at this area and sweeps throughout the length of the cardiac tube, it picks up the incoming blood and forces it out at the other end of the heart into the vessels by which it will be distributed to the body. One would have great difficulty postulating a more efficient type of pumping action in a simple tubular heart without valves.

While the heart has been building up an effective beat, blood corpuscles have been forming in the blood islands of the yolk-sac (Fig. 58). At the same time, also, the vessels from the area vasculosa to the heart have been formed, so there is an open path for the corpuscles to enter the heart. Just before the

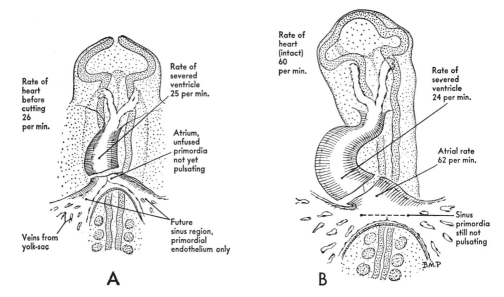

Fig. 66. Diagrams showing the location of cuts made in living hearts of young embryos. (*A*) Embryos of 10 to 12 somites. At this stage only the ventricle is pulsating, and cutting it away from the nonactive atrium has no appreciable effect on its rate. (*B*) Embryos of 13 to 15 somites. By this stage the atrium has begun to pulsate and, in the intact heart, acts as a pacemaker. Transection between atrium and ventricle shows the atrium maintaining essentially the rate of the intact heart, whereas the ventricle drops back to approximately the same slow rate it exhibited before the atrium became active and drove the ventricle at its own faster rate. [*From Patten, Univ. of Michigan Med. Bull.,* **22** (1956).]

actual circulation of blood begins, some of these corpuscles can be seen in the afferent vessels floating in fluid, and shuttling back and forth with each heartbeat. The last links of the chain to be forged are the arterial channels from the dorsal aortae out to the yolk-sac. At about the 16- to 17-somite stage (38 to 40 hours of incubation), these vessels open all the way out to the meshwork of small channels on the yolk-sac which is dotted with blood islands. If one is watching a living embryo when the final channels open, one sees the shuttling of corpuscles in the afferent vessels near the heart give way to a jerky progression. This dramatic series of events in the beginning of the circulation of blood can be described with such sureness as to details because it has been watched step by step in living embryos, and micro-moving-picture records have been made of all the critical phases of the process (Patten and Kramer, 1933).

If we turn our attention now to the routes followed by the blood that has started to circulate, the functional significance of the entire system should be apparent. In tracing the course of either the embryonic or the vitelline circulation the heart is the logical starting point. From the heart the blood of the extra-embryonic vitelline circulation passes through the ventral aortic roots, thence through the aortic arches and along the dorsal aortae, and out through the omphalomesenteric arteries to the plexus of vessels on the yolk (Fig. 67). In the small vessels which ramify in the membranes enveloping the yolk the blood absorbs food material. In young embryos, before the allantoic circulation has appeared, the vitelline circulation is involved also in the oxygenation of the blood. The great surface exposure

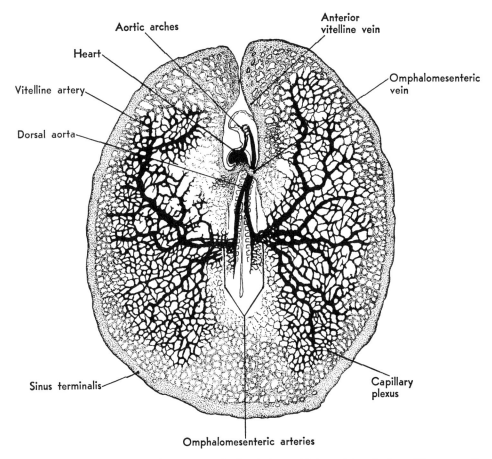

Fig. 67. The vitelline circulation in a chick of about 44 hours' incubation. Diagrammatic ventral view based on Popoff's figures of injected embryos. The arteries are shown in solid black; the veins are stippled. Note the rich plexus of small, freely anastomosing vessels in the splanchnopleure of the yolk-sac.

presented by the multitude of small vessels on the yolk makes it possible for the blood to take up oxygen which penetrates the porous shell and the albumen.

After acquiring food material and oxygen the blood is collected by the sinus terminalis and the vitelline veins. The vitelline veins converge toward the embryo from all parts of the vascular area and empty into the omphalomesenteric veins which return the blood to the heart (Fig. 67).

The blood of the intra-embryonic circulation, leaving the heart, enters the ventral aortic roots, thence passes by way of the aortic arches into the dorsal aortae, and is distributed through branches from the dorsal aortae to the body of the embryo. It is returned from the cephalic part of the body by the anterior cardinals, and from the caudal part of the body by the posterior cardinals. The anterior and posterior cardinals discharge together through the common cardinal veins into the sino-atrial region of the heart (Fig. 64).

In the heart, the blood of the extra-embryonic circulation and of the intra-embryonic circulation is mixed. The mixed blood in the heart is not so rich in oxygen and food material as that which comes to the heart from the vitelline circulation nor so low in food and oxygen content as that returned to the heart from the intra-embryonic circulation, where these materials are drawn upon by the growing tissues of the embryo. Nevertheless it carries a sufficient proportion of food and oxygen so that as it is distributed to the body of the embryo it serves to supply the growing tissues.

CHAPTER 10

EXTRA-EMBRYONIC MEMBRANES

The Folding Off of the Body of the Embryo   In bird embryos the somatopleure and splanchnopleure extend over the yolk peripherally, beyond the region where the body of the embryo is being formed. Distal to the body of the embryo the layers are termed extra-embryonic. At first the body of the chick has no definite boundaries, and consequently embryonic and extra-embryonic layers are directly continuous without there being any definite boundary at which we may say one ends and the other begins. As the body of the embryo takes form, folds develop about it, undercut it, and finally nearly separate it from the yolk. The folds which thus definitely establish the boundaries between intra-embryonic and extra-embryonic regions are known as the limiting body folds or simply the *body folds*.

The first of the body folds to appear is the fold which marks the boundary of the head. By the end of the first day of incubation the head has grown anteriorly, and the fold originally bounding it appears to have undercut and separated it anteriorly from the blastoderm (Figs. 2 and 47). The cephalic limiting fold at this stage is crescentic, concave caudally. As this fold continues to progress caudad, its posterior extremities become continuous with folds which develop along either side of the embryo. Because of the fact that these folds bound the body of the embryo laterally, they are known as the lateral body folds *(lateral limiting sulci)*. The lateral body folds, at first shallow (Fig. 63, *D*), become deeper, undercutting the body of the embryo from either side and further separating it from the yolk (Fig. 68).

During the third day a fold appears bounding the posterior region of the embryo. This caudal fold undercuts the tail of the embryo forming a subcaudal pocket (Fig. 69, *C*) just as the subcephalic fold undercuts the head. The combined effect of the development of the subcephalic, lateral body, and the subcaudal folds is to constrict off the embryo more and more from the yolk (Figs. 68 and 71). These folds which establish the contour of the embryo indicate at the same time the boundary between the tissues, which are built into the body of the embryo, and the so-called extra-embryonic tissues, which serve protective and nutritive functions during development but are not incorporated in the structure of the adult body.

The Establishing of the Yolk-sac and the Delimitation of the Embryonic Gut
The extra-embryonic membranes of the chick are four in number: the yolk-sac, the amnion, the serosa, and the allantois. The yolk-sac is the first of these to make its appearance. The splanchnopleure of the chick, instead of forming a closed gut as happens in forms with little yolk, grows over the yolk surface. The primitive gut has a cellular wall dorsally only, with the yolk acting as a temporary floor (Fig. 69, *A*). The extra-embryonic extension of the splanchnopleure eventually forms a sac-like investment for the yolk (Figs. 68 and 71).

Concomitantly with the spreading of the extra-embryonic splanchnopleure about the yolk, the intra-embryonic splanchnopleure is undergoing a series of changes which result in the establishment of a completely walled gut in the body of the embryo. The interrelations of the various steps in the formation of the gut and of the yolk-sac make it necessary to repeat some points and anticipate other points concerning the formation of the gut, in order that their relation to yolk-sac formation may not be overlooked.

It will be recalled that the first part of the primitive gut to acquire a cellular floor is its cephalic region. The same folding process by which the head is separated from the blastoderm involves the entoderm of the gut. The part of the primitive gut which acquires a floor as the subcephalic fold progresses caudad is termed the foregut (Fig. 69, *B*). During the third day of incubation the caudal fold undercuts the posterior end of the embryo. The splanchnopleure of the gut is involved in the progress of the subcaudal fold, so that a hindgut is established in a manner similar to that involved in the formation of the foregut (Fig. 69, *C*). The part of the gut which still remains open to the yolk is known as the midgut. As the embryo is constricted off from the yolk by the progress of the subcephalic and subcaudal folds, the foregut and hindgut are increased in extent at the expense of the midgut. The midgut is finally diminished until it opens ventrally by a small aperture which flares out, like an inverted funnel, into the yolk-sac (Fig 69, *D*). This opening is the yolk-duct, and its wall constitutes the yolk-stalk.

The walls of the yolk-sac are still continuous with the walls of the gut along the constricted yolk-stalk thus formed, but the boundary between the intra-embryonic splanchnopleure of the gut and the extra-embryonic splanchnopleure of the yolk-sac can now be established definitely at the yolk-stalk.

As the neck of the yolk-sac is constricted, the omphalomesenteric arteries and omphalomesenteric veins, caught in the same series of foldings, are brought together and traverse the yolk-stalk side by side. The vascular network in the splanchnopleure of the yolk-sac, which in young chicks was seen spreading over the yolk, eventually nearly encompasses it. The embryo's store of food material thus comes to be suspended from the gut of the midbody region in a sac provided with a circulatory arc of its own, the *vitelline arc*. Apparently no yolk passes directly through the yolk-duct into the intestine. Absorption of the yolk is effected through the activity of the entodermal cells lining the yolk-sac. These cells produce digestive enzymes which change the yolk into soluble form. This soluble material can then be absorbed through the lining of the yolk-sac and passed on through the endothelial lining of the vitelline blood vessels to the circulating blood by which it is carried to all parts of the growing embryo (Fig. 100). In older embryos (Fig. 68, *C*, *D*) the epithelium of the yolk-sac undergoes a series of foldings which greatly increase its surface area and thereby the amount of absorption it can accomplish.

During development the albumen loses water, becomes more viscid, and rapidly decreases in bulk. The growth of the allantois, an extra-embryonic structure which we have yet to consider, forces the albumen toward the distal end of the yolk-sac (Fig. 68, *D*). The details of the manner in which the albumen is encompassed between the yolk-sac and folds of the allantois and serosa belong to later stages of development than those with which we are concerned. Suffice it to say that the albumen, like the yolk, is eventually encompassed by the yolk-sac by which it is absorbed and transferred, by way of the extra-embryonic circulation, to the embryo.

Toward the end of the period of incubation, usually on the 19th day, the remains of the yolk-sac are enclosed within the body walls of the embryo.

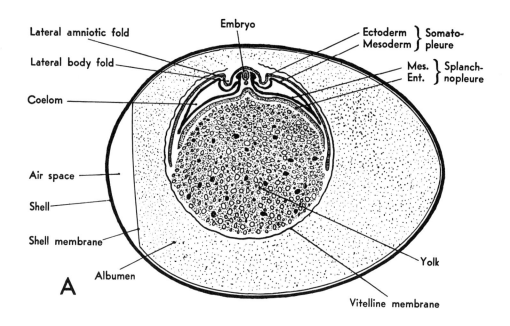

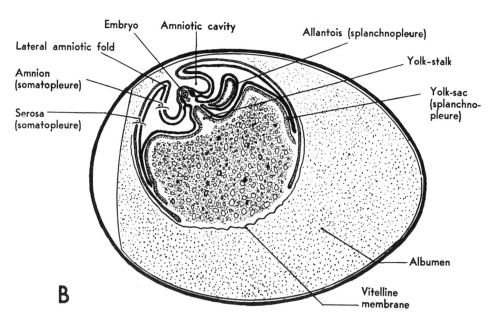

Fig. 68. Schematic diagrams to show the extra-embryonic membranes of the chick. The diagrams represent longitudinal sections through the entire egg. The body of the embryo, being oriented approximately at right angles to the long axis of the egg, is cut transversely. (*A*) Em-

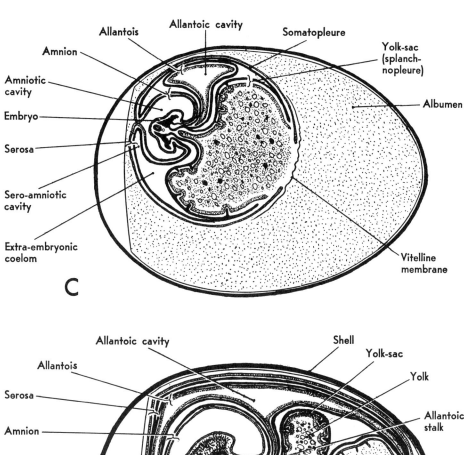

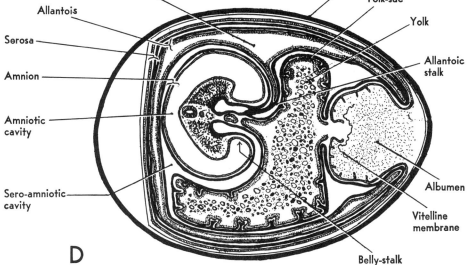

bryo of about 2 days' incubation. *(B)* Embryo of about 3 days' incubation. *(C)* Embryo of about 5 days' incubation. *(D)* Embryo of about 14 days' incubation. *(Based on Duval.)*

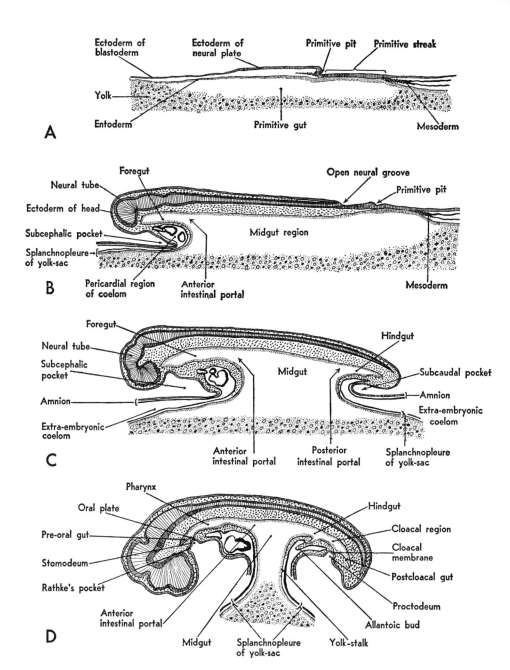

Fig. 69. Schematic longitudinal-section diagrams of the chick, showing four stages in the formation of the gut tract. The embryos are represented as unaffected by torsion. (*A*) Chick toward the end of the first day of incubation; no regional differentiation of primitive gut is as yet apparent. (*B*) Toward the end of the second day; foregut established. (*C*) Chick of about 2½ days; foregut, midgut, and hindgut established. (*D*) Chick of about 3½ days; foregut and hindgut increased in length at expense of midgut; yolk-stalk formed.

After its inclusion in the embryo, both the wall and the remaining contents of the yolk-sac rapidly disappear, their absorption being practically completed in the first six days after hatching.

The Amnion and the Serosa   The amnion and the serosa are so closely associated in their origin that they must be considered together. Both are derived from the extra-embryonic somatopleure. The amnion encloses the embryo as a saccular investment, and the cavity thus formed between the amnion and the embryo becomes filled with a watery fluid. Suspended in this amniotic fluid, the embryo is free to change its shape and position, and external pressure upon it is equalized. Smooth muscle fibres develop in the amnion, which by their contraction gently agitate the amniotic fluid. The slow and gentle rocking movement thus imparted to the embryo apparently aids in keeping its growing parts free from one another, thereby preventing adhesions and resultant malformations.

The first indication of amnion formation appears in chicks of about 30 hours incubation. The head of the embryo sinks into the yolk somewhat, and at the same time the extra-embryonic somatopleure anterior to the head is thrown into a fold, the head fold of the amnion (Fig. 71, *A*). In dorsal aspect the margin of this fold is crescentic in shape with its concavity directed toward the head of the embryo. As the embryo increases in length, its head grows anteriorly into the amniotic fold. At the same time growth in the somatopleure itself tends to extend the amniotic fold caudad over the head of the embryo (Fig. 71, *B*). By continuation of these two growth processes the head soon comes to lie in a double-walled pocket of extra-embryonic somatopleure which covers the head like a cap (Fig. 70). The free edge of the amniotic pocket retains its original crescentic shape as, in its progress caudad, it covers more and more of the embryo (Figs. 65 and 74).

The caudally directed limbs of the head fold of the amnion are continued posteriorly along either side of the embryo as the lateral amniotic folds. The lateral folds of the amnion grow dorso-mesiad, eventually meeting in the midline dorsal to the embryo (Fig. 68, *A–C*).

During the third day, the tail fold of the amnion develops about the caudal region of the embryo. Its manner of development is similar to that of the head fold of the amnion, but it grows in the opposite direction, its concavity being directed anteriorly and its progression being cephalad (Fig. 71, *B, C*).

Continued growth of the head, lateral, and tail folds of the amnion results in their meeting above the embryo. At the point where the folds meet, they become fused in a scar-like thickening termed the *amniotic raphe*

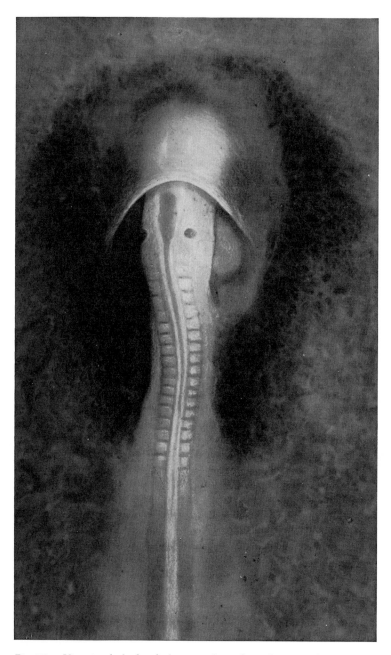

Fig. 70. Unstained chick of about 40 hours' incubation, photographed by reflected light to show the cephalic fold of the amnion enveloping the head of the embryo.

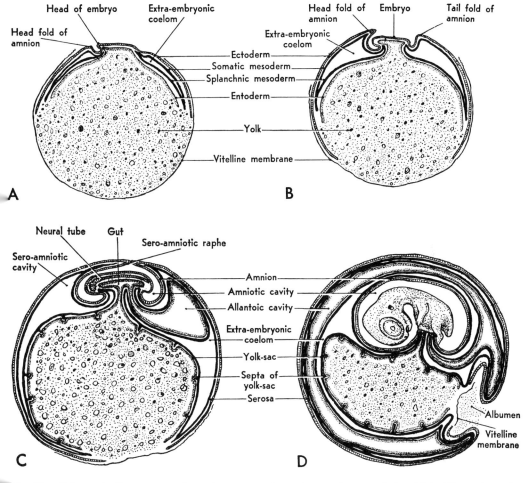

Fig. 71. Schematic diagrams to show the extra-embryonic membranes of the chick. The embryo is cut longitudinally. The albumen, shell membranes, and shell are not shown; for their relations see Fig. 68. (*A*) Embryo early in second day of incubation. (*B*) Embryo early in third day of incubation. (*C*) Embryo of 5 days. (*D*) Embryo of 9 days. (*D, after Lillie.*)

or (sero-amniotic raphe) (Fig. 71, *C*). The way in which the somatopleure has been folded about the embryo leaves the amniotic cavity completely lined by ectoderm, which is continuous with the superficial ectoderm of the embryo at the region where the yolk-stalk enters the body (Fig. 68, *D*).

All the amniotic folds involve doubling the somatopleure on itself. Only the inner layer of the somatopleuric fold, however, is directly involved in the formation of the amniotic cavity. The outer layer of somatopleure becomes the serosa (Fig. 68, *B*). The cavity between serosa and amnion (sero-amniotic cavity) is part of the extra-embryonic coelom. The continuity of the extra-embryonic coelom with the intra-embryonic coelom is most apparent in early stages (Fig. 68, *A* and *B*). They remain, however, in open communication in the yolk-stalk region until relatively late in development.

The rapid peripheral growth of the somatopleure carries the serosa about the yolk-sac, which it eventually envelops. The albumen-sac also is surrounded by folds of serosa, and the allantois, after its establishment, develops within the serosa, between it and the amnion (Fig. 73). Thus the serosa eventually encompasses the embryo itself and all the other extra-embryonic membranes (Figs. 68, *D*, and 71, *D*). The raltionships of the serosa and allantois and the functional significance of the serosa will be taken up after the allantois has been considered.

The Allantois   The allantois differs from the amnion and serosa in that it arises within the body of the embryo (Fig. 73, *A*). Its proximal portion remains intra-embryonic throughout development. Its distal portion, however, is carried outside the confines of the intra-embryonic coelom and becomes associated with the other extra-embryonic membranes (Fig. 73, *B, C*). Like the other extra-embryonic membranes, the distal portion of the allantois functions only during the incubation period and is not incorporated into the structure of the adult body.

The allantois first appears late in the third day of incubation. It arises as a diverticulum from the ventral wall of the hindgut, and its walls are, therefore, splanchnopleure. Its relationships to structures within the embryo will be better understood when chicks of 3 and 4 days incubation have been studied, but its general location can be appreciated from the schematic diagrams of Figs. 71 and 73.

During the fourth day of development the allantois pushes out of the body of the embryo into the extra-embryonic coelom. Its proximal portion lies parallel to the yolk-stalk and just caudal to it. When the distal portion of the allantois has grown clear of the embryo, it becomes enlarged (Fig. 71, *C*). Its narrow proximal portion is known as the allantoic stalk, the enlarged distal portion as the allantoic vesicle. Fluid accumulating in the allantois distends it so the appearance of its terminal portion in entire embryos is somewhat balloon-like (Fig. 72).

The allantoic vesicle enlarges very rapidly from the fourth to the tenth day of incubation. Extending into the sero-amniotic cavity it becomes flattened and finally encompasses the embryo and the yolk-sac (Figs. 68, *C, D*, and 71, *C, D*). In this process the mesodermic layer of the allantois becomes fused with the adjacent mesodermic layer of the serosa. There is

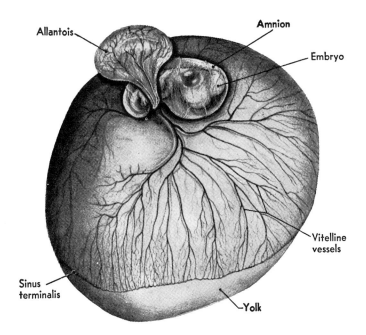

Allantois

Amnion

Embryo

Vitelline
vessels

Sinus
terminalis

Yolk

Fig. 72.   Chick of about 5½ days' incubation taken out of the shell
with yolk intact. The albumen and the serosa have been removed
to expose the embryo lying within the amnion. The allantois has
been displaced upward in order to show the relations of the allantoic stalk. Compare this figure with Fig. 68, C, which shows schematically the relations of the membranes in a section through an
embryo of similar age. (*Modified from Kerr.*)

thus formed a double layer of mesoderm, the serosal component of which
is somatic mesoderm and the allantoic component of which is splanchnic
mesoderm. In this double layer of mesoderm an extremely rich vascular network develops which is connected with the embryonic circulation by the
allantoic arteries and veins. It is through this circulation that the allantois
carries on its primary function of oxygenating the blood of the embryo and
relieving it of carbon dioxide. This is made possible by the position occupied
by the allantois, close beneath the porous shell (Figs. 68 and 98). It is this
highly vascular fusion membrane, commonly called the *chorio-allantoic
membrane*, that has been used so effectively as a site for grafting small explants from younger embryos for the purpose of testing their developmental
potencies.

    In addition to the respiratory interchange of oxygen and carbon di-
oxide, the growth of the embryo, of course, involves the metabolism of pro-

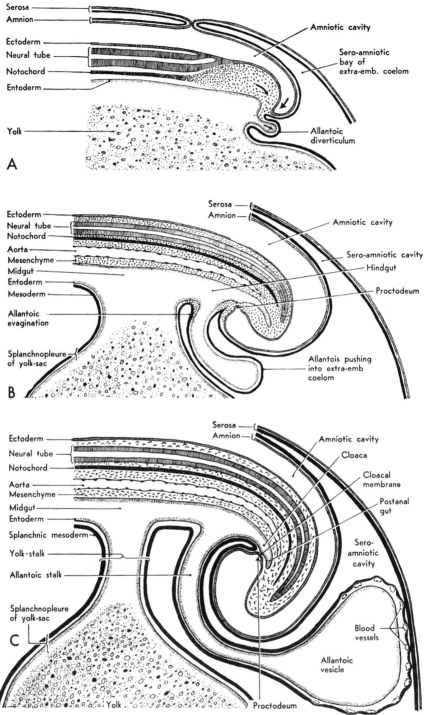

Fig. 73. Schematic longitudinal-section diagrams of the caudal regions of a series of chick embryos to show the formation of the allantois. (*A*) At about 2¾ days of incubation; (*B*) at about 3 days; (*C*) at 4 days.

teins with the formation of urea and uric acid. The averting of toxic effects from the accumulation of these waste products in an embryo growing within the confines of a shell presents some interesting problems. The allantois is again involved, for it serves as a reservoir for the secretions coming from the developing excretory organs (Fig. 98). In the early stages of development the chick excretes mostly urea. Later the excreted material becomes chiefly uric acid. This is a significant change, for urea is a relatively soluble substance and requires large amounts of water to hold it at nontoxic levels. It can be cared for while the embryo is young and its excretory output small, but would present grave problems if it were produced in large quantities. In contrast uric acid is relatively insoluble, and the large amounts of it produced by older embryos can be stored without ill effects. At the time of hatching the slender allantoic stalk is broken, and the distal portion of the allantois with its contained excretory products remains as a shriveled membrane adherent to the broken shell.

The fusion of the allantoic mesoderm and blood vessels with the serosa is of particular interest because of its homology with the establishment of the chorion in the higher mammals.[1] The chorion of mammalian embryos arises by the fusion of allantoic vessels and mesoderm with the inner wall of the serosa and constitutes the embryo's organ of attachment to the uterine wall. In mammalian embryos the allantoic, or umbilical, circulation, as it is usually called in mammals, serves more than a respiratory function. In the absence of any appreciable amount of yolk, the mammalian embryo derives its nutrition, through the allantoic circulation, from the uterine blood of the mother. Thus the mammalian allantoic circulation carries out the functions which in the chick are divided between the vitelline and the allanotic circulations.

[1] The serosa of the chick is occasionally called chorion. The term chorion, however, is more frequently applied to the composite layer formed by the fusion of the allantois and the serosa. The latter is a much more proper usage, since these two conjoined layers, rather than the serosa alone, correspond with the chorion of mammals. In primates the homologies are somewhat masked by running together of the early steps in the process, and by reducation of size of the allantoic cavity; yet even the human chorion can be analyzed into an outer portion homologous with the serosa of the chick and an inner portion (intimately fused to the outer from their first appearance), which represents allantoic vessels and mesoderm.

In some books the term outer or false amnion will be found used to designate the structure in this book called serosa. The term false amnion is not, however, in general use in this country.

CHAPTER 11

STRUCTURE OF CHICKS FROM FIFTY TO FIFTY-FIVE HOURS OF INCUBATION

I.   EXTERNAL FEATURES

In chicks which have been incubated from 50 to 55 hours (Fig. 74) the entire head has been freed from the yolk by the progress caudad of the subcephalic fold. Torsion has involved the whole anterior half of the embryo and is completed in the cephalic region, so that the head now lies left side down on the yolk. The posterior half of the embryo is still in its original position, ventral surface prone on the yolk. At the extreme posterior end, the beginning of the caudal fold marks off the tail region of the embryo from the extra-embryonic membranes. The head fold of the amnion has progressed caudad, together with the lateral amniotic folds, impocketing the embryo nearly to the level of the omphalomesenteric arteries.

The *cranial flexure*, which was seen beginning in chicks of about 38 hours, has increased rapidly until at this stage the brain is bent nearly double on itself. The axis of the bending being in the midbrain region, the mesencephalon comes to be the most anteriorly located part of the head, and the prosencephalon and myelencephalon lie opposite each other, ventral surface facing ventral surface (Fig. 74). The original anterior end of the prosencepha-

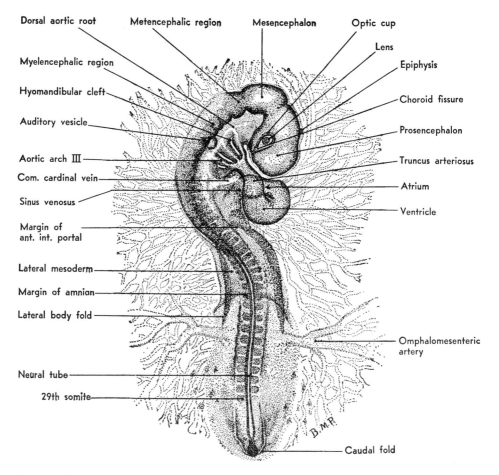

Fig. 74. Dextro-dorsal view (×17) of entire embryo of 29 somites (about 55 hours' incubation).

lon is thus brought in close proximity to the heart, and the optic vesicles and the auditory vesicles are brought opposite each other at nearly the same antero-posterior level.

At this stage flexion has involved the body farther caudally as well as in the brain region. It is especially marked at about the level of the heart in the region of transition from myelencephalon to spinal cord. Since this is the future neck region of the embryo, the flexure at this level is known as the *cervical flexure.*

II. THE NERVOUS SYSTEM

Growth of the Telencephalic Region   The completion of flexion and torsion in the cephalic region is accompanied by marked changes in the configuration of the brain. The same fundamental regions can, however, be identified throughout this range of development. In embryos of 50 hours (Fig. 75) a slight indentation in the dorsal wall of the prosencephalon indicates the impending division of this part of the brain into telencephalon and diencephalon. By 60 hours this demarcation has become more definite (Fig. 78). Except for its considerable increase in size, no important changes have as yet taken place in the telencephalic region.

The Epiphysis   In the middorsal wall of the diencephalic region a small evagination has appeared. This evagination is the epiphysis (Figs. 74 and 75). It is destined to become differentiated into the pineal body of the adult.

The Infundibulum and Rathke's Pocket   In the floor of the diencephalon the infundibular depression has become deepened and lies close to a newly formed ectodermal invagination known as Rathke's pocket (Fig. 75). The epithelium of Rathke's pocket is destined to be separated from the superficial ectoderm and to become permanently associated with the infundibular portion of the diencephalon to form the *hypophysis* or *pituitary gland.*

The Optic Vesicles   The optic vesicles have undergone changes which completely alter their appearance. In 33-hour chicks they are hemispheroidal vesicles opening off the lateral walls of the prosencephalon (Fig. 55). At this stage the lumen of each optic vesicle *(opticoele)* is widely continuous with the lumen of the prosencephalon *(prosocoele)* (Fig. 63, *A*). The constriction

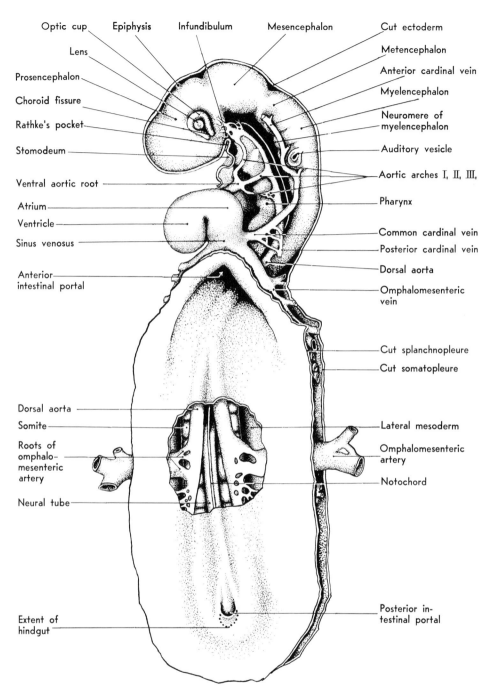

Optic cup | Epiphysis | Infundibulum | Mesencephalon | Cut ectoderm

Lens — Metencephalon

Prosencephalon — Anterior cardinal vein

Choroid fissure — Myelencephalon

Rathke's pocket — Neuromere of myelencephalon

Stomodeum — Auditory vesicle

Ventral aortic root — Aortic arches I, II, III,

Atrium — Pharynx

Ventricle —

Sinus venosus — Common cardinal vein

Posterior cardinal vein

Dorsal aorta

Anterior intestinal portal — Omphalomesenteric vein

Cut splanchnopleure

Cut somatopleure

Dorsal aorta —

Somite — Lateral mesoderm

Roots of omphalo-mesenteric artery — Omphalomesenteric artery

Neural tube — Notochord

Extent of hindgut — Posterior intestinal portal

Fig. 75. Diagram of dissection of chick of about 50 hours. The splanchnopleure of the yolk-sac cephalic to the anterior intestinal portal, the ectoderm of the left side of the head, and the mesoderm in the pericardial region have been dissected away. A window has been cut in the splanchnopleure of the dorsal wall of the midgut to show the origin of the omphalomesenteric arteries. (*Modified from Prentiss.*)

of the optic stalk which begins to be apparent in 38-hour embryos (Fig. 56) is much more marked in 55-hour chicks.

The most striking and important advance in their development is the invagination of the distal ends of the single-walled optic vesicles to form double-walled *optic cups* (Fig. 76, *B*). The concavities of the cups are directed laterally. Mesially the cups are continuous, over the narrowed *optic stalks,* with the ventro-lateral walls of the diencephalic region of the original prosencephalon. The invaginated layer of the optic cup is termed the *sensory layer* because it is destined to give rise to the sensory layer of the retina. The layer against which the sensory layer comes to lie after its invagination is termed the *pigment layer* because it gives rise to the pigmented layer of the retina. The double-walled cups formed by invagination are also termed *secondary optic vesicles* in distinction to primary optic vesicles, as they are called before their invagination. The formerly capacious lumen of the primary optic vesicle is practically obliterated in the formation of the optic cup. What remains of the primary opticoele is now but a narrow space between the sensory and the pigment layers of the retina (Fig. 76, *B*). Later when these two layers fuse, this space is entirely obliterated.

While the secondary optic vesicles are usually spoken of as the optic cups, they are not complete cups. The invagination which gives rise to the secondary optic vesicles, instead of beginning at the most lateral point in the primary optic vesicles, begins at a point somewhat toward their ventral surface and is directed mesio-dorsad. As a result the optic cups are formed without any lip on their ventral aspect. They may be likened to cups with a segment broken out of one side. This gap in the optic cup is the *choroid fissure* (Fig. 75). In figure 76, *B*, a section is shown which passes through the head of the embryo on a slight slant so that the right optic cup, being cut to one side of the choroid fissure, appears complete; while the left optic cup, being cut in the region of the fissure, shows no ventral lip.

The infolding process by which the optic cups are formed from the primary optic vesicles is continued into the region of the optic stalks. As a result their ventral surfaces become grooved. Later in development the optic nerves and blood vessels come to lie in the grooves thus formed in the optic stalks.

The Lens   The lens of the eye arises from the superficial ectoderm of the head adjacent to the optic vesicles. The first indications of lens formation in chicks appear at about 40 hours of incubation as local thickenings of the ectoderm immediately overlying the optic vesicles. These thickenings, called *lens placodes,* sink below the general level of the surface of the head to

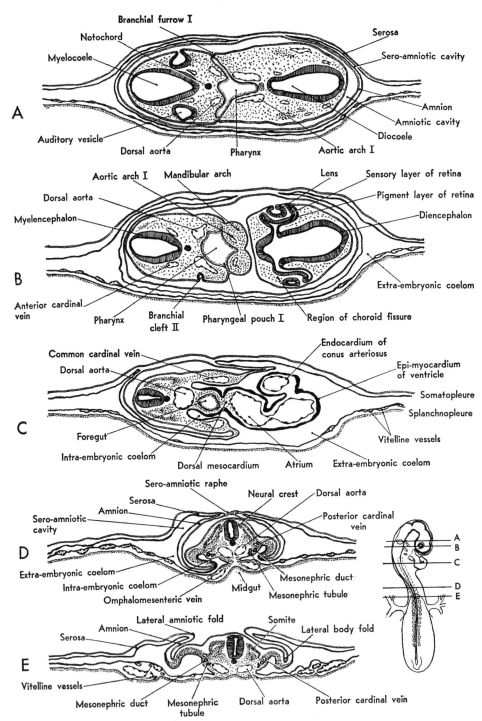

Fig. 76. Diagrams of transverse sections of 55-hour (30-somite) chick. The locations of the sections are indicated on an outline sketch of the entire embryo.

form small vesicles which extend into the secondary optic vesicles. Their opening to the surface is rapidly constricted, and eventually they are disconnected altogether from the superficial ectoderm. At this stage the opening to the outside still persists, although it is very small (Fig. 76, *B*, right eye). In sections which do not pass directly through the opening, the lens vesicle appears as if it were completely separated from the overlying ectoderm (Fig. 76, *B*, left eye).

The derivation of the lens from a placode of thickened epithelium which sinks below the general surface, and eventually loses its connection with the superficial ectoderm, is strikingly similar to the early steps in the derivation of the auditory vesicle. But these primordia, after they are separated from the ectoderm, follow divergent lines of differentiation leading to adult conditions which are structurally and functionally totally unlike. The origin of these two structures from cell groups similarly folded off from the same germ layer, but which once established undergo each their own characteristic differentiation, exemplifies a sequence of events so characteristic of developmental processes in general as to call for at least a comment in passing.

Much experimental work has been done, especially with amphibian embryos, on the influence of the optic cup on the formation of the lens. It is apparent from these studies that lens formation is induced by the optic cup in a manner comparable to the induction of neural plate formation by the underlying notochordal tissue. If the optic vesicles are removed early in development, no lens forms from the overlying ectoderm. When a young optic vesicle is transplanted to an ectopic position, such as under the ectoderm of the back, it will induce the formation of a lens vesicle from ectoderm which does not ordinarily so differentiate. If an optic vesicle is carefully removed from its normal location at a time just before the lens placode becomes morphologically differentiated, the overlying ectoderm will go on and form a lens vesicle. This means that the inductive influence has already been operative before we can see visible evidences of a response. If, now, the same young optic cup which had been effective in inducing lens formation in the normal location is transplanted to an ectopic position, it may still be capable of inducing the formation of an extra lens from the overlying ectoderm. The time in which such transplantations are effective in producing a second lens by the inductive effect of the same optic cup is, however, very sharply limited. If the optic cup is moved to its ectopic location after the formation of the normal lens is well under way, there is no response from the ectoderm overlying it in its new position. This means that the inductive effect is operative at only a certain limited period of development. If anything interferes with its action at that time it can never again exert its directive influence.

This makes it evident that there is a factor of "timing" in developmental processes that is of more far-reaching significance than is generally recognized. Disturbances in the timing of one developmental process with reference to another are undoubtedly involved in the genesis of many types of developmental abnormalities. For an embryo to develop normally it is necessary not only that all the essential growing materials be available in the right quantities and at the right places, but also that they must be there at the right time. We can make the crude comparison of a growing embryo with a building contractor, whose plans call for the laying of steel-reinforced concrete at a certain place at a certain time. If the cement is mixed, ready to pour, and the truck bringing in the steel reinforcing goes into the ditch in transit, the cement may harden and become unpourable before the necessary steel arrives.

The Posterior Part of the Brain and the Spinal Cord Region of the Neural Tube   Caudal to the diencephalon the brain shows no great change as compared with the last stages considered. The mesencephalon is somewhat enlarged and the constrictions separating it from the diencephalon cephalically, and the metencephalon caudally, are more sharply marked. The metencephalon is more clearly marked off from the myelencephalon, and its roof is beginning to show thickening. In the myelencephalon the neuromeric constrictions are still evident in the ventral and lateral walls (Figs. 74 and 75). The dorsal wall has become much thinner than the ventral and lateral walls (Fig. 76, A, B) and shows no trace of division between the neuromeres.

In the spinal cord region of the neural tube the lateral walls have become thickened at the expense of the lumen so that the neural canal appears slit-like in sections of embryos of this age (Fig. 76, E) rather than elliptical as it is immediately after the closure of the neural folds. At this stage the closure of the neural tube is completed throughout its entire length. The last regions to close were at the cephalic and caudal ends of the neural groove. In younger stages where they remained open these regions were known as the anterior neuropore and the sinus rhomboidalis, respectively.

The Neural Crest   In the closure of the neural tube the superficial ectoderm which at first lay on either side of the neural groove, continuous with the neural plate ectoderm, becomes fused in the midline and separated from the neural plate to constitute an unbroken ectodermal covering (cf. Figs. 50, B, and 63, B). At the same time the lateral margins of the neural plate become fused to complete the neural tube. There are cells lying originally at the edges of the neural folds which are not involved in the fusion of either the superficial ectoderm or the neural plate. These cells form a pair of longi- tudinal aggregations extending one on either side of the middorsal line in the angles between the superficial ectoderm and the neural tube (Fig. 77, A). With the fusion of the edges of the neural folds to complete the neural tube, and the fusion of the superficial ectoderm dorsal to the neural tube, these two longitudinal cell masses become for a time confluent in the midline (Fig. 77, B). But it should be emphasized that this aggregation of cells arises from paired components and soon again separates into right and left parts. On account of its temporary position dorsal to the neural tube it is known as the *neural crest.*

The neural crest should not be confused with the margin of the neural fold with which it is associated before the closure of the neural tube. The margin of the neural fold involves cells which go into the superficial ecto- derm and into the neural tube, as well as those which are concerned in the formation of the neural crest.

When first established, the neural crest is continuous antero-posteri- orly. As development proceeds, the cells of the neural crest migrate ventro- laterally on either side of the spinal cord (Fig. 77, C) and at the same time become segmentally clustered. The segmentally arranged cell groups thus derived from the neural crest give rise to the dorsal root ganglia of the spinal nerves, and in the cephalic region to the ganglia of the sensory cranial nerves. (For a later stage of the dorsal root ganglia see Fig. 92.)

III.   THE DIGESTIVE TRACT

The Foregut   The manner in which the three primary regions of the gut tract are established has already been considered in a general way. (See Chap. 10 and Fig. 69.) In 50- to 55-hour chicks the foregut has acquired considerable length. It extends from the anterior intestinal portal cephalad almost to the region of the infundibulum (Fig. 75).

As the first part of the tract to be established, the foregut is naturally the most advanced in differentiation. We can already recognize a pharyngeal and an esophageal portion. The pharyngeal region lies ventral to the mye- lencephalon and is encircled by the aortic arches (Fig. 78). The pharynx is somewhat flattened dorso-ventrally and has a considerably larger lumen than the esophageal part of the foregut (cf. Fig. 76, B and C).

The Stomodaeum   There is at this stage no mouth opening into the phar- ynx. However, the location where the opening will be formed is indicated

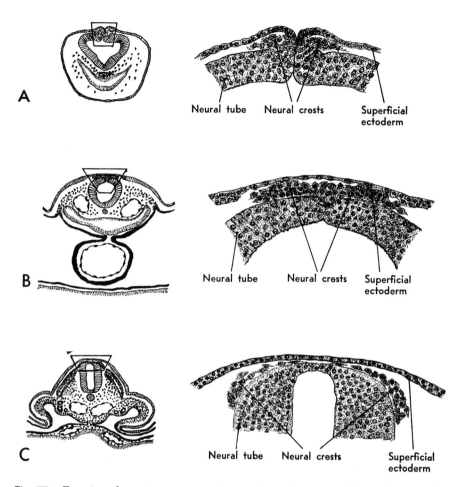

Fig. 77. Drawings from transverse sections to show the origin of neural crest cells. The location of the area drawn is indicated on the small sketch to the left of each drawing. (*A*) Anterior rhombencephalic region of 30-hour chick. (*B*) Posterior rhombencephalic region of 36-hour chick. (*C*) Middorsal region of cord in 55-hour chick.

by the approximation of a ventral outpocketing near the anterior end of the pharynx to a depression formed in the adjacent ectoderm of the ventral surface of the head (Fig. 75). The ectodermal depression, known as the *stomodaeum*, deepens until its floor lies in contact with the entoderm of the pharyngeal outpocketing (Fig. 75). The thin layer of tissue formed by the apposition of the stomodaeal ectoderm to the pharyngeal entoderm is known as the *oral plate*. Later in development the oral plate breaks through bringing the stomodaeum and the pharynx into open communication and thereby establishing the oral opening.

The Pre-oral Gut   It will be noted by reference to Fig. 75 that the oral opening is not established at the extreme cephalic end of the pharynx. The part of the pharynx which extends cephalic to the mouth opening is known as the *pre-oral gut*. After the rupture of the oral plate, the pre-oral gut eventually disappears, but an indication of it persists for a time as a small diverticulum termed Seessel's pocket (cf. Figs. 75 and 89).

The Midgut   Although the midgut is still the most extensive of the three primary divisions of the digestive tract, it presents little of interest. It is nothing more than a region where the gut still lies open to the yolk. It does not have even a fixed identity. As fast as any part of the midgut acquires a ventral wall by the closing-in process involved in the progress of the sub-cephalic and subcaudal folds, it ceases to be midgut and becomes foregut or hindgut. Differentiation and local specializations appear in the digestive tract only in regions which have ceased to be midgut.

The Hindgut   The hindgut first appears in embryos of about 50 hours. The method of its formation is similar to that by which the foregut was established. The subcaudal fold undercuts the tail region and walls off a gut pocket the entrance into which is the *posterior intestinal portal* (Figs. 69, C, and 75). The hindgut is lengthened at the expense of the midgut as the sub-caudal fold progresses cephalad and is also lengthened by its own growth caudad. It shows no local specializations until later in development.

IV.   THE BRANCHIAL (GILL) CLEFTS AND BRANCHIAL (GILL) ARCHES

At this stage the chick embryo has unmistakable branchial (gill) arches and branchial (gill) clefts. Although only transitory, they are morphologically of great importance not only from the comparative viewpoint and because of their significance as structures exemplifying recapitulation, but also because of their involvement in the formation of the embryonic arterial system, some of the ductless glands, the Eustachian tube, and the face and jaws.

The *branchial clefts* are formed by the meeting of ectodermal depressions, the *branchial furrows*, with diverticula from the lateral walls of the pharynx, the *pharyngeal pouches*. During most of the time the branchial furrows are conspicuous features in entire embryos, they may be seen, by studying sections, to be closed by a thin layer of tissue composed of the ectoderm of the floor of the branchial furrow and the entoderm at the distal extremity of the pharyngeal pouch (Fig. 76, *A*). The breaking through of this thin double layer of tissue brings the pharyngeal pouches into communication with the branchial furrows, thereby establishing open branchial clefts. In birds an open condition of the clefts is transitory. In the chick the most posterior of the series of clefts never becomes open. Although some of the clefts never become open and others open for a short time, the term cleft is usually used to designate these structures which, whether open or not, are clearly homologous with the gill clefts of water-living ancestral forms.

The position of the branchial or gill clefts is best seen in entire embryos. They are commonly designated by number beginning with the first cleft posterior to the mouth and proceeding caudad. The first post-oral cleft appears earliest in development and is discernible at about 46 hours of incubation. Branchial cleft II appears soon after, and by 50 to 55 hours three clefts have been formed (Fig. 74).

Between adjacent branchial clefts, the lateral body-walls about the pharynx are thickened. Each of these lateral thickenings meets and merges in the midventral line with the corresponding thickening of the opposite side of the body. Thus the pharynx is encompassed laterally and ventrally by a series of arch-like thickenings, the branchial, or gill, arches. The branchial arches, like the branchial clefts, are designated by number, beginning at the anterior end of the series. Branchial arch I lies cephalic to the first post-oral cleft, between it and the mouth region. Because of the part it plays in the formation of the mandible, it is also designated as the *mandibular arch*. Branchial arch II is frequently termed the *hyoid arch*, and visceral cleft I, because of its position between the mandibular and hyoid arches, is known as the *hyomandibular cleft*. Posterior to the hyoid arch the branchial arches and clefts are ordinarily designated by their post-oral numbers only.

There are other structures which are just beginning to be differentiated in the pharyngeal region and foregut of embryos of this stage, but it seems better to consider them in connection with later stages when their significance will be more readily grasped.

V.   THE CIRCULATORY SYSTEM

The Heart   In embryos of 30 to 40 hours incubation we traced the expansion of the heart till it was bent to the right of the embryo in the form of a U-shaped tube (Figs. 53, 55, and 57). The disappearance of the dorsal meso-

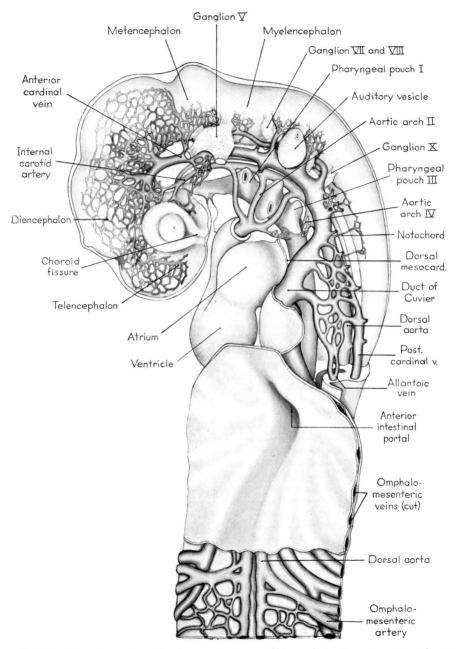

Ganglion Ⅴ

Metencephalon

Myelencephalon

Ganglion Ⅶ and Ⅷ

Pharyngeal pouch I

Anterior
cardinal
vein

Auditory vesicle

Aortic arch Ⅱ

Internal
carotid
artery

Ganglion Ⅹ

Pharyngeal
pouch Ⅲ

Aortic
arch Ⅳ

Diencephalon

Notochord

Dorsal
mesocard.

Choroid
fissure

Duct of
Cuvier

Dorsal
aorta

Telencephalon

Post.
cardinal v.

Atrium

Allantoic
vein

Ventricle

Anterior
intestinal
portal

Omphalo-
mesenteric
veins (cut)

Dorsal aorta

Omphalo-
mesenteric
artery

Fig. 78. Drawing to show the deeper structures of the cephalo-thoracic region of a 60-hour chick, exposed from the left. The basis of the illustration was a wax-plate reconstruction made from serial sections; the smaller vessels and the primordial capillary plexuses were added from injected specimens. Note especially the relations of the aortic arches to the pharyngeal pouches and the way in which the large veins enter the sinus venosus. Although the omphalomesenteric veins are not fully exposed by this dissection, the bulges they cause in the entoderm on either side of the anterior intestinal portal clearly suggest the way the main right and left veins become confluent with each other to enter the sinus venosus as a short median trunk. The old term for the common cardinal vein (duct of Cuvier) has been retained in the labeling of this figure.

cardium, except at its posterior end leaves the midregion of the heart lying unattached and extending to the right, into the pericardial region of the coelom. The heart is fixed with reference to the body of the embryo at its cephalic end, where the ventral aortic roots lie embedded beneath the floor of the pharynx, and caudally in the sino-atrial region, where it is attached by the omphalomesenteric veins, the common cardinal veins, and the persistent portion of the dorsal mesocardium.

During the period between 30 and 55 hours of incubation the heart itself is growing more rapidly than is the body of the embryo in the region where the heart lies. Since its cephalic and caudal ends are fixed, the unattached midregion of the heart must undergo bending. It becomes at first U-shaped and then twisted on itself to form a loop. The atrial region of the heart is forced somewhat to the left, and the truncus is thrown across the atrium by being twisted to the right and dorsally. The ventricular region constitutes the loop proper (cf. Figs. 56, 65, and 74). This twisting process reverses the original cephalo-caudal relations of the atrial and ventricular regions. Before the twisting, the atrial region of the heart was caudal to the ventricular region as it is in the adult fish heart. In the twisting of the heart the atrial region, by reason of its association with the fixed sinus region of the heart, undergoes relatively little change in position. The ventricular region is carried over the dextral side of the atrium and comes to lie caudal to it, thus arriving in the relative position it occupies in the adult heart of birds and mammals.

The bending and subsequent twisting of the heart lead toward its division into separate chambers. As yet, however, no indication of the actual partitioning of the heart into right and left sides is apparent. It is still essentially a tubular organ through which the blood passes directly, without any division into separate channels.

The Aortic Arches   In 33- to 38-hour chicks the ventral aortae communicate with the dorsal aortae over a single pair of aortic arches which bend around the anterior end of the pharynx (Figs. 57 and 64). With the formation of the branchial arches new aortic arches appear. The original pair of aortic arches comes to lie in the mandibular arch (Fig. 79). New aortic arches are formed caudal to the first pair, one pair in each branchial arch. In chicks of 50 hours, two pairs of aortic arches have been established (Fig. 80), and frequently the third also is present or starting to form (Fig. 75). Usually by 60 hours the fourth aortic arch is beginning to take shape (Fig. 78).

The Fusion of the Dorsal Aortae   The dorsal aortae arise as vessels paired throughout their entire length (Fig. 57). As development progresses

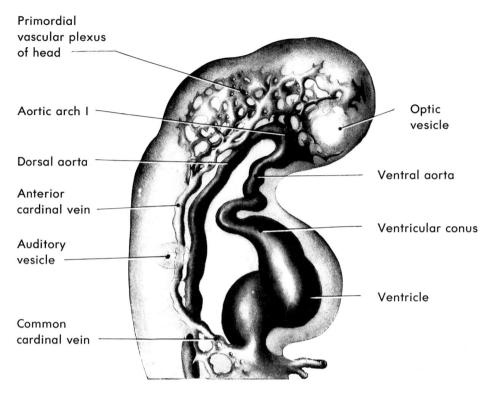

Primordial
vascular plexus
of head

Aortic arch I

Dorsal aorta

Anterior
cardinal vein

Auditory
vesicle

Common
cardinal vein

Optic
vesicle

Ventral aorta

Ventricular conus

Ventricle

Fig. 79.   Dextral view of cephalic and cardiac regions of chick of about 45 hours' incubation. The blood vessels have been injected to show the capillary plexus of the cephalic region. In the drawing the heart and arteries are differentiated from the veins and capillaries by darker shading. This figure, together with Fig. 80, shows the manner in which main vessels develop in the embryo from a primary capillary plexus. It will be noted in this figure that the part of the capillary plexus which lies more superficially already shows an enlargement in diameter and a directness of path which is indicative of the fact that it is to become the main trunk vein of this region. (*From Minot after Evans.*)

they fuse in the midline to form the unpaired dorsal aorta familiar in adult anatomy. This fusion takes place first at cardiac level (Fig. 76, C). Cephalically it never extends to the pharyngeal region. Caudally the whole length of the aorta is eventually involved. By 55 hours of incubation the fusion has progressed caudad to about the level of the fourteenth somite.

The Cardinal and Omphalomesenteric Vessels   The relationships of the cardinal veins and the omphalomesenteric vessels are little changed from the conditions seen in 40- to 50-hour chicks. The posterior cardinals have elongated, keeping pace with the caudal progress of differentiation in the mesoderm. They lie just dorsal to the intermediate mesoderm in the angle formed between it and the somites (Figs. 76, *D*, and 81, *C*). As the omphalomesenteric veins approach the sino-atrial part of the heart, they come to lie progressively closer to each other in the midline. It is at this region of the convergence of the omphalomesenteric veins that the common cardinal veins enter them dorso-laterally (Fig. 78). This region of confluence of the great veins, as we shall see in considering older stages, is being involved in the moulding of the sinus venosus. The omphalomesenteric arteries, meanwhile, are tending to lose the multiple roots by which they emerged from the dorsal

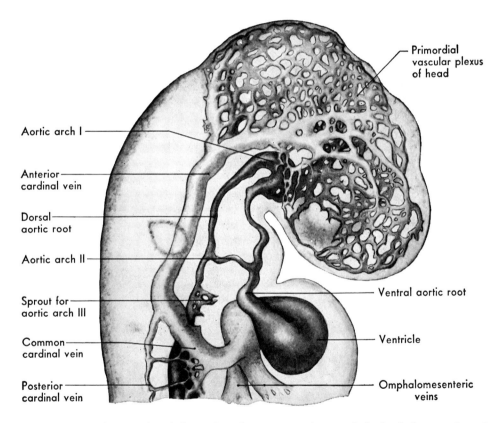

Fig. 80. Dextral view of cephalic and cardiac regions of injected chick of about 50 hours' incubation. This figure shows a later stage in the development of the anterior cardinal vein from the primary capillary plexus of the cephalic region. The main channel, only vaguely suggested in the previous figure, is here quite definite. Here also a second aortic arch has been completed, and a plexiform outgrowth of vascular endothelium from the dorsal aortic root toward the ventral aorta indicates the impending formation of the third aortic arch. (*From Minot after Evans, redrawn with modifications in the cardiac region.*)

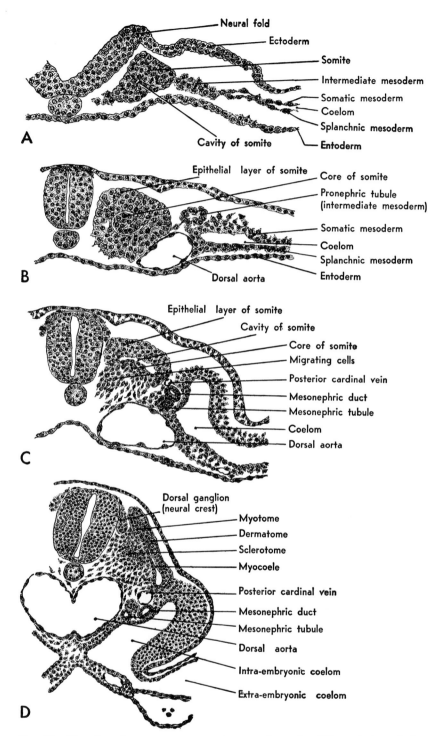

Fig. 81. Drawings from transverse sections to show the differentiation of the somites. (*A*) Second somite of a 4-somite chick; (*B*) ninth somite of a 12-somite chick; (*C*) twentieth somite of a 30-somite chick; (*D*) seventeenth somite of a 33-somite chick.

aortae in younger stages, but otherwise they exhibit essentially the same relationships.

VI.   THE DIFFERENTIATION OF THE SOMITES

When the somites are first formed, they consist of nearly solid masses of cells derived from the dorsal mesoderm (Fig. 81, *A*). The cells composing them show a more or less radial arrangement. In the center of the somite a cavity is usually discernible. This cavity is at first extremely minute, and in somites which have been recently formed it may be altogether wanting.

As a somite becomes more sharply marked off, the radial arrangement of the outer zone of its cells becomes more definite (Fig. 81, *B*). The boundaries of the central cavity are considerably extended, but its lumen is almost completely filled by a core of irregularly arranged cells. In sections which pass through the middle of the somite, this central core of cells is seen to arise from the lateral wall of the somite where it is continuous with the intermediate mesoderm.

A little later in development the outer zone of cells on the ventro-mesial face of the somite loses its originally definite boundaries and becomes merged with the central core of cells. This ill-defined cell aggregation, known as the *sclerotome*, becomes mesenchymal in characteristics, and extends ventro-mesiad from the somite of either side toward the notochord (Fig. 81, *C, D*). The cells of the sclerotomes of either side continue to converge about the notochord. Later in development they will take part in the formation of the axial skeleton.

During the emergence of sclerotomal cells, the dorso-lateral part of the original outer cell-zone of the somite has maintained its definite boundaries and epithelioid characteristics. The part of this outer zone which thus lies parallel to the ectoderm is known as the *dermatome* (Fig. 81, *D*). It received this name because its cells were believed to migrate out and come to underlie the ectoderm, giving rise to the connective tissue layer (dermis) of the integument. Although some cells from this region of the somite undoubtedly are contributed to the formation of the deep layers of the skin, the conviction has been gaining ground that many, perhaps most, of them take part in the formation of muscle. Furthermore, the connective tissue layer of the skin is known to receive many cells from the somatic mesoderm generally, and from the diffuse mesenchyme in the cephalic region where there are no somites. The term dermatome is so firmly fixed that it is probably unwise to attempt to discard it, but we should bear in mind that although it does contribute to the dermis, it probably does not do so any more extensively than other regions of the mesoderm which lie in close proximity to the ectoderm.

The dorso-mesial portion of the outer zone of the somite becomes the *myotome*. It is folded somewhat laterad from its original position next to the neural tube (Fig. 81, *C*) and comes to lie ventro-mesial to the dermatome and parallel to it (Fig. 81, *D*). (A later stage in the differentiation of the somite is shown in Fig. 92.) The portion of the original cavity which persists for a time between the dermatome and myotome is termed the *myocoele*. The myotomes undergo the most extensive growth of any of the parts of the somite, giving rise eventually to the major part of the skeletal musculature of the body.

VII. THE URINARY SYSTEM

In the section diagrams of Fig. 76, *D, E*, certain parts of the urinary system which have been established in chicks of 50 to 55 hours will be found located and labeled. The urinary system is relatively late in becoming differentiated, and only a few of the early steps in its formation can at this time be made out. Many structures which later become of great importance are not represented even by primordial cell aggregations. Except for those who are well grounded in comparative anatomy, any logical discussion of the structures which have appeared must anticipate much that occurs later in development. Consideration of the mode of origin and significance of the nephric organs appearing at this stage has, therefore, been deferred.

CHAPTER 12

DEVELOPMENT OF THE CHICK DURING THE THIRD AND FOURTH DAYS OF INCUBATION

I.   EXTERNAL FEATURES

Torsion   Chicks of 3 days' incubation (Fig. 82) have been affected by torsion throughout their entire length. Torsion is complete well posterior to the level of the heart, but the caudal portion of the embryo is not yet completely turned on its side. In 4-day chicks the entire body has been turned through 90 degrees, and the embryo lies with its left side on the yolk (Figs. 72 and 84).

Flexion   The cranial and cervical flexures which appeared in embryos during the second day have increased so that in 3-day and 4-day chicks the

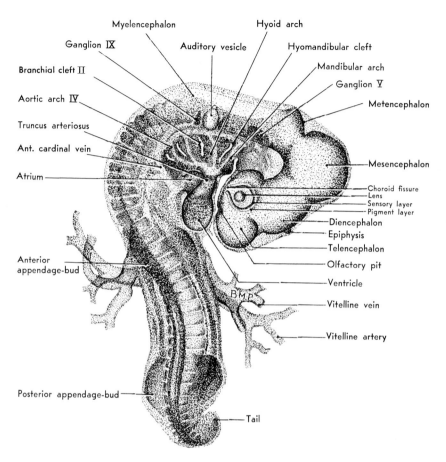

Fig. 82. Dextro-dorsal view (×16) of entire chick embryo of 36 somites (about 3 days' incubation).

long axis of the embryo shows nearly right-angled bends in the midbrain and in the cervical region. The midbody region of 3-day chicks is slightly concave dorsally. This is due to the fact that the embryo is still broadly attached to the yolk in that region. By the end of the fourth day the body folds have undercut the embryo, so it remains attached to the yolk only by a slender stalk. The yolk-stalk soon becomes elongated, allowing the embryo to become first straight in the middorsal region and then convex dorsally. At the same time the caudal flexure is becoming more pronounced. The progressive increase in the cranial, cervical, dorsal, and caudal flexures results in the bending of the embryo on itself so that its originally straight long axis becomes C-shaped, and its head and tail lie close together (Figs. 83 and 84).

buds have increased considerably in size to form paddle-shaped extensions from the sides of the body (Fig. 83). The main mass of the buds is composed of closely packed mesenchymal cells; the outer covering is of ectoderm. When seen in sections, there is a conspicuously thickened band of this ecto-dermal covering along the convex outer margin of the buds. This is the *apical ectodermal ridge* (Figs. 87, *B*, and 97, *E*). It is nearly half through the incuba-tion period before the appendages take on avian characteristics (Fig. 86).

There are exceedingly interesting interactions between the mesodermal core and the ectodermal apical ridge of a developing appendage-bud. The earliest indication of impending limb formation is to be seen in chicks around the middle of the third day of incubation. There is then a heightened rate of cell proliferation in the somatic layer of the lateral mesoderm at the levels where appendage-buds are destined to appear. In these regions cells emerge from the parent layer and become mesenchymal in their characteristics (Fig. 87, *A*). It is these mesenchymal cells which, migrating laterad, constitute the core of the appendage-bud. Moreover they induce the formation of the apical ridge in the overlying ectoderm (Fig. 87, *B*). We have already con-sidered various examples of such inductive activity in other locations. Of particular interest in this instance is the way in which, once established, the ectodermal apical ridge exercises a return controlling action on the mesen-chymal core of the bud. If the apical ridge is removed, the distal part of the appendicular skeleton fails to develop. Wing bones fail to form in the anter-ior buds (Fig. 87, *C–F*) or foot bones in the posterior bud. The earlier in development the apical ridge is removed the greater is the difficiency in the development of the distal parts of the appendage. It should be emphasized that such disturbances are limited to the more distal portions of the limbs and that the appendicular girdles are not affected.

Another interesting feature of the development of the appendages is the way in which some of their molding is brought about by selective cell death. By using such vital dyes as Nile blue sulfate or neutral red, Saunders and his coworkers (1962) have shown areas of cell deterioration to be ex-tensively involved in the shaping of digits or wing parts in the original pad-dle-shaped appendage-buds. We have been accustomed to thinking of de-

---

tion. Note that by this stage of development definitely avian features have emerged from an early embryonic configuration which was much the same as that of other sauropsidan embryos. (See Fig. 1.) The enormous relative size of the eye and the midbrain; the shape of the appendage-buds, tail, and beak; the auditory meatus with no ear pinna; the nodules on crown, back, and tail due to growing feather germs—all unmistakably place the embryo as belonging to the bird class.

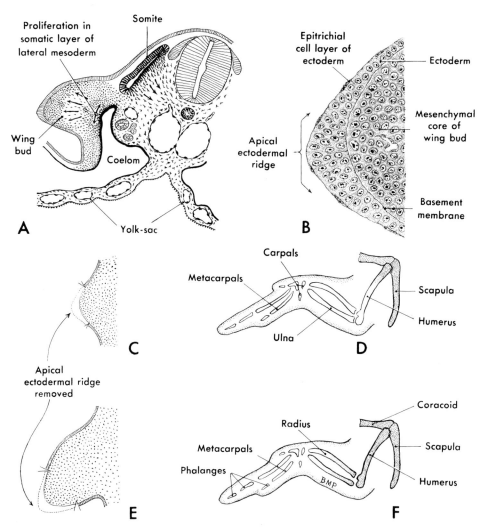

Fig. 87. The effect of the removal of the apical ectodermal ridge on the development of the wing-bud. (*A*) Diagram showing location of the center of mesodermal proliferation involved in forming the core of the wing-bud. (*B*) Cell detail drawing of tip of wing-bud of 3-day chick to show the apical ridge. (*C*) Ridge removal in 3-day chick. (*D*) Skeletal deficiency resulting from third day removal of apical ridge. (*E*) Ridge removal at 4 days. (*F*) Skeletal deficiences resulting from fourth day removal of apical ridge. In *D* and *F* the parts of the skeleton which develop are stippled; the parts failing to develop are shown in outline only. [*C to F, showing the results of apical ridge removal on skeletal development, based on the work of Saunders, J. Exptl. Zool.,* **108** (1948).]

velopmental processes almost exclusively in terms of growth activities. The concept of selective cell death as a molding process adds a new dimension to our thinking in regard to morphogenetic processes.

II.   THE NERVOUS SYSTEM

Summary of Development Prior to the Third Day   The earliest indication of the formation of the central nervous system appears in chicks of 16 to 18 hours as a local thickening of the ectoderm which forms the neural plate (Fig. 38). The neural plate then becomes longitudinally folded to form the neural groove (Figs. 43, 45, 47, and 48). By fusion of the margins of the neural folds, first in the cephalic region and later caudally, the neural groove is closed to form a tube and at the same time separated from the body ectoderm. The cephalic portion of the neural tube becomes dilated to form the brain, and the remainder of the neural tube gives rise to the spinal cord (Figs. 52 and 55).

In its early stages the brain shows a series of enlargements in its ventral and lateral walls, indicative of its fundamental metameric structure. In the establishment of the three-vesicle condition of the brain, the lines of demarcation between prosencephalon, mesencephalon, and rhombencephalon are formed by the exaggeration of certain of the inter-neuromeric constrictions and the obliteration of others. (See Chap. 8 and Fig. 54.) The original neuromeric enlargements persist longest in the rhombencephalon.

The three-vesicle condition of the brain is transitory. By 40 hours the division of the rhombencephalon into metencephalon and myelencephalon is clearly indicated (Figs. 54, *D,* and 56). The division of the prosencephalon and the establishment of the five-vesicle condition characteristic of the adult brain do not take place until somewhat later.

In chicks of 55 hours (Figs. 74 and 75) the development of the cranial flexure has resulted in the bending of the brain so that the entire prosencephalon is displaced first ventrad and then caudad toward the heart. At the same time the head of the embryo has undergone torsion and lies with its left side on the yolk. Although flexion and torsion have thus completely changed the general relationships of the brain as seen in entire embryos, the regions already established in 40-hour chicks are still evident. The prosencephalon has, however, become very noticeably enlarged cephalic to the optic vesicles, and a slight constriction in its dorsal wall indicates the beginning of the demarcation of the telencephalic region from the diencephalic region.

Formation of the Telencephalic Vesicles   By the end of the third day the antero-lateral walls of the primary forebrain have been evaginated to form a pair of vesicles lying one on either side of the midline (Fig. 82). These lateral evaginations are known as the *telencephalic vesicles*. In chicks of 4 days the telencephalic vesicles have undergone considerable enlargement (Figs. 83, 85, and 88). The openings through which their cavities are continuous with the

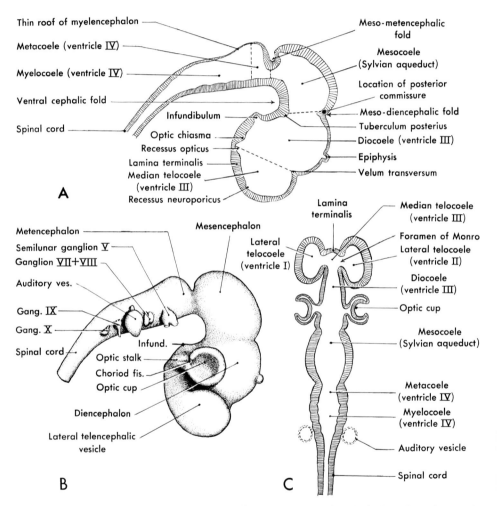

Fig. 88.   Diagrams to show the topography of the brain of a 4-day chick. (*A*) Plan of sagittal section. The arbitrary boundaries between the various brain vesicles (according to von Kupffer) are indicated by broken lines. (*B*) Dextral view of a brain which has been dissected free. (*C*) Schematic frontal section plan of brain. The flexures of the brain are supposed to have been straightened before the section was cut.

lumen of the median portion of the brain are later known as the *foramina of Monro*. The telencephalic division of the brain includes not only these two lateral vesicles but also the median portion of the brain from which they arise. The lumen of the telencephalon has therefore three divisions, a *median telocoele*, broadly confluent posteriorly with the diocoele, and two lateral telencephalic vesicles, connecting with the median telocoele through the foramina of Monro (Fig. 88, C).

Before the formation of the telencephalic vesicles the most anterior part of the brain lay in the midline, but the rapid growth of the telencephalic vesicles soon carries them rostrally beyond the median portion of the telocoele. The median anterior wall of the telocoele which formerly was the most rostral part of the brain, and which remains the most rostral part of the brain lying in the midline, is known as the *lamina terminalis* (Fig. 88, A, C). The telencephalic vesicles become the cerebral hemispheres, and their cavities become the paired lateral ventricles of the adult brain. The hemispheres undergo enormous enlargement in their later development and extend dorsally and posteriorly as well as rostrally, eventually covering the entire diencephalon and mesencephalon under their posterior lobes. They contain the brain-centers for memory and for all actions conditioned by past experience.

As a matter of convenience in dealing with the morphology of the brain, more or less arbitrary lines of division between the adjacent brain regions are recognized. The division between telencephalon and diencephalon is an imaginary line drawn from the velum transversum to the recessus opticus (Fig. 88, A). *Velum transversum* is the name given to the internal ridge formed by the deepening of the dorsal constriction which was first noted in chicks of 55 hours as indicating the impending division of the primary forebrain (Fig. 75). The *recessus opticus* is a transverse furrow in the floor of the brain which in the embryo leads on either side into the lumina of the optic stalks. Because it is located just rostral to the optic chiasma where some of the optic nerve fibers cross, it is often spoken of as the *preoptic recess*.

The Diencephalon   The lateral walls of the diencephalon at this stage show little differentiation except ventrally where the optic stalks merge into the walls of the brain. The development of the epiphysis as a median evagination in the roof of the diencephalon has already been mentioned (Chap. 11). Except for some enlargement, it does not differ from its condition when first formed in embryos of about 55 hours. The infundibular depression in the floor of the diencephalon has become appreciably deepened and lies in close

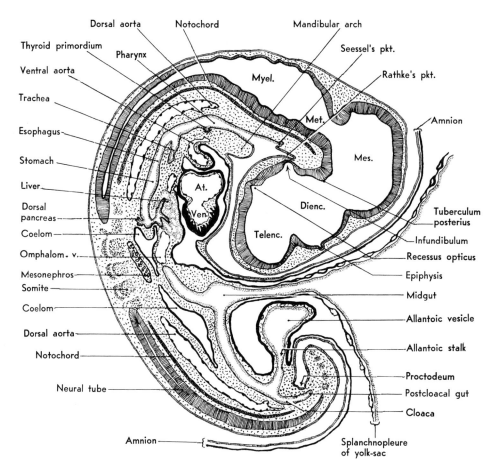

Fig. 89.   Diagram of median longitudinal section of 4-day chick. Because of a slight bend in the embryo the section is parasagittal in the middorsal region, but for the most part it passes through the embryo in the sagittal plane.

proximity to Rathke's pocket (Fig. 89) with which it is destined to fuse in the formation of the hypophysis. Later in development the lateral walls of the diencephalon become greatly thickened to form the thalami, thus reducing the size and changing the shape of the diocoele, which is known in adult anatomy as the third brain ventricle. The anterior part of the roof of the diencephalon remains thin and becomes richly vascular. Later these vessels, invaginating the roof with them, push into the third ventricle to form the anterior choroid plexus.

The boundary between the diencephalon and the mesencephalon is an imaginary line drawn from the internal ridge formed by the original dorsal constriction between the primary forebrain and midbrain, to the tuberculum posterius (Fig. 88, *A*). The *tuberculum posterius* is a rounded elevation in the floor of the brain, of importance chiefly because it is regarded as marking the boundary between diencephalon and mesencephalon.

The Mesencephalon   The mesencephalon as yet shows no specializations, beyond a thickening of its walls. The dorsal walls of the mesencephalon later increase rapidly in thickness and become the *corpora quadrigemina* of the adult brain. As the name implies, these are four symmetrically placed elevations. The anterior pair (*superior colliculi*) constitute the brain-center for visual reflexes; the posterior pair (*inferior colliculi*) are the center for auditory reflexes. The floor of the mesencephalon also becomes greatly thickened and is known in the adult as the *crura cerebri*. It serves as the main pathway of the fiber tracts which connect the cerebral hemispheres with the posterior part of the brain and the spinal cord. The originally capacious mesocoele is thus reduced by the thickening of the walls about it to a narrow canal, *the cerebral aqueduct* or *aqueduct of Sylvius*.

The Metencephalon   The boundary between the mesencephalon and metencephalon is indicated by the original inter-neuromeric constriction which separated them at the time of their establishement (cf. Figs. 54 and 88). The caudal boundary of the metencephalon is not definitely defined. It is regarded as being located approximately at the point where the brain roof changes from the thickened condition characteristic of the metencephalon to the thin condition characteristic of the myelencephalon. The metencephalon shows practically no differentiation in 4-day chicks. Later there is ventrally and laterally an extensive ingrowth of fiber tracts giving rise to the pons and to the cerebellar peduncles. The roof of the metencephalon undergoes extensive enlargement and becomes the cerebellum of the adult brain, the coordinating center for complex muscular movements.

The Myelencephalon   The dorsal myelencephalic wall is reduced in thickness, indicative of its final fate as the thin roof of the medulla. It later receives a rich supply of small blood vessels which, carrying the roof with them, grow into the myelocoele to form the posterior choroid plexus. The ventral and lateral myelencephalic walls become the floor and side walls of the medulla. Functionally the medulla serves both as a conduction path between cord and brain and as a reflex center for involuntary activities such as breathing.

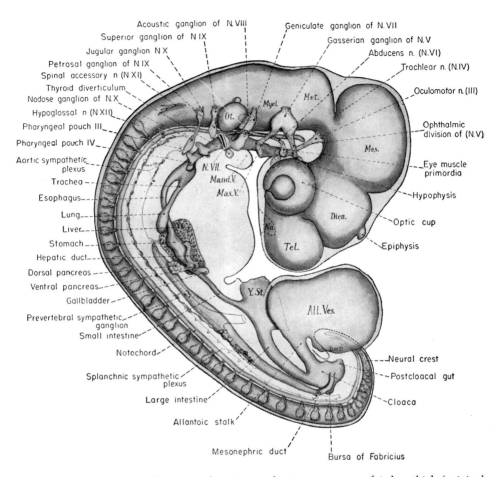

Acoustic ganglion of N.VIII
Geniculate ganglion of N.VII
Superior ganglion of N IX
Gasserian ganglion of N.V
Jugular ganglion N X
Abducens n. (N.VI)
Petrosal ganglion of N IX
Trochlear n. (N.IV)
Spinal accessory n (N XI)
Thyroid diverticulum
Oculomotor n.(III)
Nodose ganglion of N X
Hypoglossal n (N XII)
Pharyngeal pouch III
Ophthalmic division of (N.V)
Pharyngeal pouch IV
Aortic sympathetic plexus
Eye muscle primordia
Trachea
Esophagus
Hypophysis
Lung
Liver
Optic cup
Stomach
Epiphysis
Hepatic duct
Dorsal pancreas
Ventral pancreas
Gallbladder
Preverterbral sympathetic ganglion
Small intestine
Notochord
Neural crest
Splanchnic sympathetic plexus
Postcloacal gut
Large intestine
Cloaca
Allantoic stalk
Mesonephric duct
Bursa of Fabricius

Myel. *Met.* *Ot.* *N.VII* *Mand.V.* *Max.V.* *Mes.* *Dien.* *Na.* *Tel.* *Y.St.* *All.Ves.*

Fig. 90. Reconstruction of nervous, digestive, and urinary systems of 4-day chick (original ×51, reproduced ×16). From the same reconstruction as the colored illustration appearing as Fig. 91, in which all the systems are shown in their relations to each other.

Abbreviations: All. Ves., allantoic vesicle; *Dien.*, diencephalon; *Mand. V*, mandibular division of cranial nerve V (trigeminal); *Max. V*, maxillary division of cranial nerve V (trigeminal); *Mes.*, mesencephalon; *Met.*, metencephalon; *Myel.*, myelencephalon; *Na.*, location of nasal pit; *N. VII*, seventh cranial nerve (facial); *Ot.*, otic vesicle; *Tel.*, lateral telencephalic vesicle; *Y. St.*, yolk-stalk.

The Ganglia of the Cranial Nerves   In the brain region, cells derived from the cephalic portion of the neural crest have become aggregated to form ganglia. The largest and the most clearly defined of the ganglia present in 4-day chicks is the *semilunar (Gasserian) ganglion* of the fifth (trigeminal) cranial nerve (Fig. 84). It lies ventro-laterally, opposite the most anterior

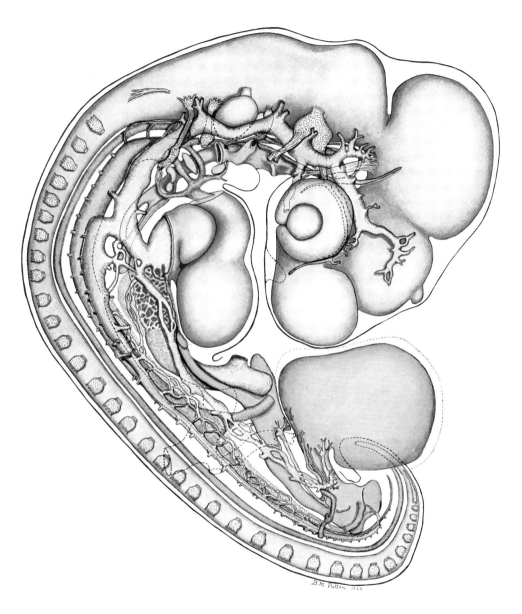

Fig. 91. Internal organs of 4-day chick. Drawn from a wax-plate reconstruction (original ×51, reproduced ×23). For keys to structures represented, see Figs. 90 and 101.

neuromere of the myelencephalon. From its cells sensory nerve fibers grow mesiad into the brain and distad to the facial and oral regions. In four-day chicks the beginning of the *ophthalmic division* of the fifth nerve extends from the ganglion toward the eye, and the beginning of the *mandibulo-maxillary division* is growing toward the angle of the mouth (Fig. 90). Immediately cephalic to the auditory vesicle is a mass of neural crest cells which is the primordium of the ganglia of the seventh and eighth nerves. The separation of this double primordium to form the geniculate ganglion of the seventh nerve and the acoustic ganglion of the eighth nerve begins during the fourth day (Fig. 90). Posterior to the auditory vesicle the superior ganglion of the ninth nerve can be clearly seen even in whole-mounts (Fig. 78). The ganglia of the tenth (vagus) nerve can be recognized in sections of chicks at the end of the fourth day but are dfficult to make out in whole-mounts. They show most clearly when reconstructed by the wax-plate method (Figs. 90 and 91).

The Spinal Cord   The spinal cord region of the neural tube when first established exhibits a lumen which is elliptical in cross section. As development progresses, the lateral walls of the cord become greatly thickened in contrast with the dorsal and ventral walls which remain thin. In this process the lumen (central canal) becomes compressed laterally until it appears in cross section as little more than a vertical slit. The thin dorsal wall of the tube is known as the *roof plate*; the thin ventral wall, as the *floor plate*; and the thickened side walls, as the *lateral plates* (Fig. 92).

The Spinal Nerves   During the fourth day the establishing of the spinal nerve roots has begun. The growth of nerve fibers from the neuroblasts can be traced only with the aid of special methods of staining. The more general steps in the development of the roots of the spinal nerves can, however, be followed in sections prepared by the ordinary methods.

In the adult each spinal nerve is connected with the cord by two roots, a *dorsal root*, which is a pathway for sensory (afferent) nerve fibers, and a *ventral root*, which is a pathway for motor (efferent) nerve fibers (Fig. 93). Lateral to the cord the dorsal and ventral roots unite. The *spinal ganglion (dorsal root ganglion)* is located on the dorsal root between the spinal cord and the point where dorsal and ventral roots unite. Distal to the union of dorsal and ventral roots is a branch, the *ramus communicans*, which extends ventrad to a ganglion of the sympathetic nerve cord (Fig. 93).

When first formed from the neural crest cells, the spinal ganglion has no connection with the cord (Fig. 77). The dorsal root is established by the

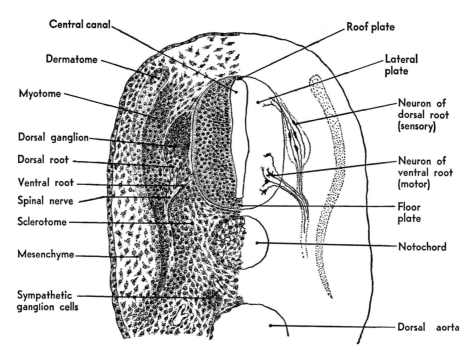

Central canal

Dermatome

Myotome

Dorsal ganglion

Dorsal root

Ventral root

Spinal nerve

Sclerotome

Mesenchyme

Sympathetic
ganglion cells

Roof plate

Lateral
plate

Neuron of
dorsal root
(sensory)

Neuron of
ventral root
(motor)

Floor
plate

Notochord

Dorsal aorta

Fig. 92. Drawing to show the structure and relations of a spinal ganglion and the roots of a spinal nerve. The left half of the drawing represents structures as they appear after treatment by the usual nuclear staining method. The right half of the section shows, schematically, the nerve cells and the fibers growing out from them as they may be demonstrated by the Golgi method. (*Nerve cells and fibers after Ramón y Cajal.*)

growth of nerve fibers from cells of the spinal ganglion mesiad into the dorsal part of the lateral plate of the cord. At the same time fibers grow distad from these cells to form the peripheral part of the nerve (Fig. 92). The fibers which arise from the dorsal root ganglion conduct sensory impulses toward the cord.

Coincident with the establishment of the dorsal root, the ventral root is formed by fibers which grow out from cells located in the ventral part of the lateral plate of the cord (Fig. 92). Most of the fibers, which thus arise from cells in the cord and pass out through the ventral root, conduct motor impulses from the brain and cord to the muscles with which they are associated peripherally.

The sympathetic ganglia arise from cells of the neural crest, and, according to some investigators, also in part from cells moving out from the spinal cord, which migrate ventrally and form masses lying on either side of the midline at the level of the dorsal aorta (Figs. 92 and 93). By the end of the fourth day the primordia of the sympathetic ganglia have become inter-connected longitudinally by slender fibrocellular strands. The developing sympathetic ganglia then appear as local enlargements on paired cord-like structures, the *prevertebral sympathetic chains* (Fig. 90). Each sympathetic ganglion is connected with the corresponding spinal nerve by a fibro-cellular cord which is the primordium of the ramus communicans. Later, both sensory and motor fibers appear in the rami communicantes, putting the sympathetic ganglia in communication with the spinal nerve roots. From certain of the sympathetic ganglia, cells migrate still farther ventrad, establishing the primordia of the splanchnic sympathetic system (Fig. 90).

The manner in which the processes of these young nerve cells grow is interesting to watch in tissue cultures. From the peripheral cytoplasm of the neuroblast a slender sprout called a *cone of growth* is formed. If this outgrowth is followed, it can be seen to advance by ameboid movements, tending to migrate along anything that furnishes a favorable substratum on which it may move. Thus the cone of growth rapidly pulls out behind itself a long slender cytoplasmic filament which is the young nerve fiber. If it encounters any resistant substance such as a bit of young cartilage in the explanted tissue or a particularly dense portion of the plasma clot, its growing tip will turn aside into paths offering less resistance.

These tendencies of a growing nerve fiber to follow along something which offers it a substratum and to be turned aside by obstacles are vitally important in dealing with the repair of nerve injuries. When a nerve is cut, the parts of the fibers which are severed from the nerve cells die. When regeneration starts, it begins with the cut ends of the parts of the fibers which remained in connection with the nerve cells. The cut ends send out cones of growth similar to those arising from embryonic neuroblasts. If the injury is a clean cut, so the proximal end of the nerve can be accurately sewed into place against the degenerated distal end, the regenerating fibers will follow along the old nerve sheaths as a pathway and eventually reestablish their peripheral connections. If, however, the approximation is not a clean one, scar tissue turns aside the regenerating nerve tips, so they do not find their way into their old pathway. If an injury is so extensive that it leaves a gap between the living ends of the nerve fibers and their former sheaths, the regenerating fibers will become hopelessly entangled in the new intervening connective tissue unless they are provided with an artificial pathway. A bit of frozen and vacuum-dried nerve may be used for this purpose. Carefully fixed across the gap in the interrupted nerve, its parallel sheaths form a path by which the regenerating nerve fibers may grow across and pick up their own old pathways. In such nerve grafts, of course, the grafted tissue serves

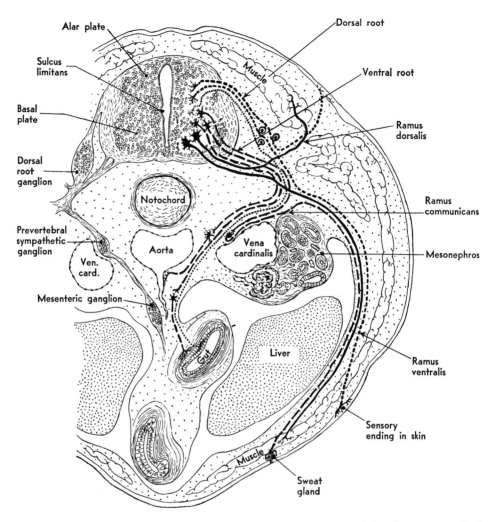

Fig. 93. Schematic diagram indicating the various connections made by the neurons which develop in a typical spinal nerve. (*Modified from Froriep.*)

The neurologist classifies the fibers in a spinal nerve according to their relations and functions. The components of a typical spinal nerve on this basis are:

I. Afferent
    A. General Somatic Afferent
        1. *Exteroceptive*, i.e., fibers conducting impulses from the external surface of the body such as touch, pain, tempera-ture. (Represented in this figure by short broken lines.)
        2. *Proprioceptive*, i.e., fibers carrying impulses of position sense from joints, tendons, and muscles. (Not represented in this diagram.)
    B. General Visceral Afferent
        Fibers from viscera (interoceptive) by way of sympathetic chain

merely as a temporary bridge and is rapidly resorbed and replaced by new host tissue.

III. THE SENSE ORGANS

The Eye The primary optic vesicles, as we have seen, arise in chicks of about 30 hours as dilations in the lateral walls of the prosencephalon (Figs. 53 and 57). At first the optic vesicles open broadly into the brain, but later constrictions develop which narrow their attachment to the form of a stalk (Fig. 56). In chicks of 55 hours the primary optic vesicles are invaginated to form the double-walled secondary optic vesicles or optic cups. The invagination takes place in such a way that the ventral wall of the cup is incomplete, the gap in it being known as the choroid fissure (Figs. 76, *B*, and 78).

ganglion, white ramus communicans, and dorsal root; cell bodies in dorsal root ganglion; no synapse before reaching cord. (Illustrated in this figure by dotted line.)

II. Efferent
  A. General Somatic Efferent
      Motor neurons to skeletal muscle; cell bodies in ventral columns of gray matter; fibers emerge by ventral roots. (Illustrated in this figure by solid lines.)
  B. General Visceral Efferent
      Two-neuron chains from cord to glands and to smooth muscle of viscera and blood vessels. The first neurons (preganglionic) have their cells of origin in the lateral column of the gray matter of cord from first thoracic to third lumbar level. Fibers leave cord by ventral root, turn off in white ramus communicans to end in synapse with the second-order neurons (postganglionic) of the two-neuron chain.

Note the various destinations of the visceral efferent paths, e.g.:
1. Fibers to smooth muscle of gut-wall; impulse relayed by second-order motor neurons from synapses in mesenteric ganglion.
2. Fibers (vasomotor) to smooth muscle of blood-vessel wall; impulses relayed by second-order motor neurons from synapses in prevertebral ganglia.
3. Fibers (pilomotor) to muscles about hair follicles, and fibers (sudomotor) to sweat glands in skin. Impulses in both these cases relayed from synapses in prevertebral ganglia by second-order motor fibers passing back over the gray ramus communicans and thence to periphery via branches of spinal nerve.

The lens, meanwhile, arises as a thickening of the superficial ectoderm which becomes depressed to form a vesicular invagination extending into the optic cup (Fig. 76, *B*).

In chicks of four days the choroid fissure has become narrowed by the growth of the walls of the optic cup on either side of it (Fig. 88, *B*). At the same time the orifice of the optic cup becomes narrowed by convergence of its margins toward the lens (Fig. 94, *A*). Meanwhile the lens has become freed from the superficial ectoderm and forms a completely closed vesicle. Sections of the lens at this stage show that the cells constituting that part of its wall which lies toward the center of the optic cup are becoming elongated to form the *lens fibers* (Fig. 94, *C*).

The Coulombres (1963) have carried out experiments on the developing lens in chick embryos, which give important information as to the factors controlling the elongation of lens fibers. With superb technical skill they removed the lens from the optic vesicle on the fifth day and replanted it with the lens epithelium facing toward the retina instead of its normal position toward the cornea. In such reversed lenses the lens fibers which had started to elongate ceased their growth, and the lens epithelium, which normally does no such thing, began to form a new set of lens fibers. This clearly indicates that the formation of lens fibers is not simply a matter of the age of the formative cells but that, taken at this early stage, the lens can respond to a changed environment by complete reversal of its polarity.

At this stage we can identify the beginning of most of the structures of the adult eye. The thickened internal layer of the optic cup will give rise to the *sensory layer of the retina* (Fig. 94, *B*). Fibers arise from nerve cells in this layer of the retina and grow along the groove in the ventral surface of the optic stalk toward the brain to form the optic nerve. The external layer of the optic cup gives rise to the *pigment layer of the retina*. Mesenchymal cells can be seen aggregating in progressively increasing numbers about the outside of the optic cup. From these the *sclera* and *choroid coat* are derived. Some of the mesenchyme makes its way into the optic cup through the choroid fissure and gives rise to the cellular elements of the *vitreous body*. The complex *ciliary apparatus* of the adult eye is derived from the margins of the optic cup adjacent to the lens. The *corneal and conjunctival epithelium* arise from the superficial ectoderm overlying the eye. Mesenchymal cells which make their way between the lens and the corneal epithelium give rise to the *substantia propria* of the cornea.

The Ear   Of the structures taking part in the formation of the ear, the first to appear is the auditory placode. The auditory placode is recognizable in

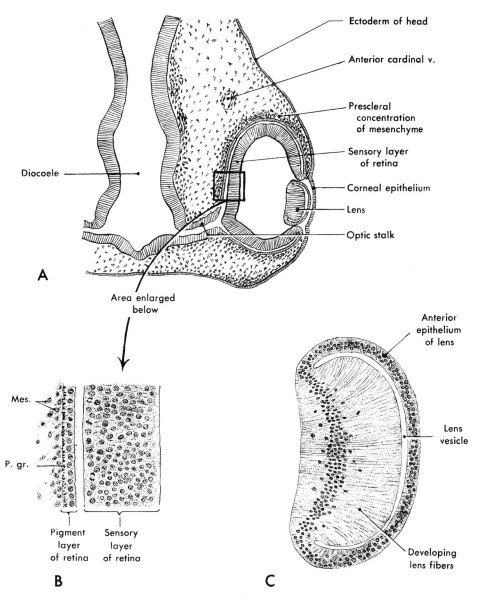

Fig. 94.   Drawings to show structure of the eye of a 4-day chick. (*A*) Diagram to show topography of eye region. (*B*) Drawing to show cellular organization of the pigment and sensory layers of the retina. (*C*) Drawing to show cellular organization of the lens.

Abbreviations: Mes., mesenchymal cell; *P. gr.*, pigment granule.

36-hour chicks as a thickened plate of ectoderm. Almost as soon as it appears, the placode sinks below the level of the surrounding ectoderm to form the floor of the auditory pit (Fig. 56). By constriction of its opening to the surface, the epithelium of the auditory pit becomes separated from the ectoderm of the head and comes to lie close to the lateral wall of the myelencephalon (Fig. 76, A). A tubular stalk, the *endolymphatic duct*, remains for a time adherent to the superficial ectoderm, marking the location of the original invagination (Fig. 83).

The degree of development reached by the ear primordium in 4-day chicks gives little indication of the nature of the later processes by which the ear is formed. The auditory vesicle by a very complex series of changes will give rise to the entire epithelial portion of the *internal ear* mechanism. Nerve fibers arising from the *acoustic ganglion* grow into the brain proximally, and to the internal ear distally, establishing nerve connections between them. There is at this stage no indication of the differentiation of the external auditory meatus. The dorsal and inner portion of the hyomandibular cleft which gives rise to the *auditory (Eustachian) tube* and to the middle ear chamber has not yet become associated with the auditory vesicle.

The Olfactory Organs   The olfactory organs are represented in 3- and 4-day chicks by a pair of depressions in the ectoderm of the head. These so-called *olfactory pits* are located ventral to the telencephalic vesicles and just anterior to the mouth (Figs. 83 and 85). By growth of the processes which surround them, the olfactory pits become greatly deepened. The epithelium lining the pits eventually comes to lie at the extreme upper part of the *nasal chambers* and constitutes the *olfactory epithelium*. Nerve fibers grow from these cells to the telencephalic lobes of the brain to form the *olfactory nerves*.

IV.   THE DIGESTIVE AND RESPIRATORY SYSTEMS

Summary of Development Prior to the Third Day   The primary entoderm which gives rise to the epithelial lining of the digestive and respiratory systems and their associated glands (Fig. 37) becomes established as a separate layer before the egg is laid (Fig. 33, A). In its early relationships the entoderm is a sheet-like layer of cells lying between the ectoderm and the yolk and attached peripherally to the yolk, so the primitive gut cavity is bounded only dorsally by the entoderm and ventrally has the yolk as a temporary floor (Fig. 69, A).

Only the part of the entoderm which lies within the embryonal area is involved in the formation of the enteric tract. The peripheral portion of the entoderm goes into the formation of the yolk-sac. There is at first no definite line of demarcation between the entoderm destined to be incorporated into the body of the embryo and that which remains extra-embryonic in its associations. The foldings which appear later, separating the body of the embryo from the yolk, establish for the first time the boundaries between intra-embryonic and extra-embryonic entoderm (Figs. 68 and 71).

The first part of the gut to acquire a complete entodermic lining is the foregut. Its floor is formed by the caudally progressing concrescence of the entoderm which takes place as the subcephalic and lateral body folds under-cut the cephalic part of the embryo (Fig. 69, B). At a considerably later stage the hindgut is formed in a similar manner by the progress of the subcaudal fold toward the head (Fig. 69, C). Between the foregut and the hindgut, the midgut remains open to the yolk ventrally. As the embryo is more completely separated from the yolk, the foregut and hindgut increase in extent at the expense of the midgut. By the fourth day of incubation the midgut is reduced to the region where the yolk-stalk opens into the enteric tract (Fig. 69, D).

Establishing of the Oral Opening   As we saw in the study of younger chicks, when the gut is first established it ends as a blind pocket both cephalically and caudally. In embryos of 55 to 60 hours the processes leading toward the establishment of the oral opening were, however, clearly indicated. A midventral evagination of the pharynx had been established immediately cephalic to the mandibular arch (Fig. 75). Opposite this outpocketing of the pharynx and growing in to meet it, the stomodaeal depression had been formed. The only bar to an actual opening was the oral plate, a thin membrane constituted by the apposition of the pharyngeal entoderm and the stomodaeal ectoderm. The communication of the foregut with the outside is finally established, during the third day, by the breaking through of the oral plate. Following the rupture of the oral plate, growth of the surrounding structures (Fig. 85) rapidly deepens the originally shallow stomodaeal depression. The region where the oral plate was originally located in the embryo eventually becomes, in the adult, the region of transition from oral cavity to pharynx.

The formation of the oral opening in the manner described does not take place at the extreme anterior end of the foregut. A small gut pocket extends cephalic to the mouth. This so-called pre-oral gut rapidly becomes less conspicuous after the breaking through of the oral plate. The small depression which in older embryos marks its location is known as Seessel's pocket (Fig. 89). Even this small depression eventually disappears altogether. Its importance lies wholly in the fact that its margin indicates for some time the place at which ectoderm and entoderm originally became continuous in the formation of the oral opening.

The Pharynx   Caudal to the oral opening the foregut has become flattened dorso-ventrally and widened laterally to form the pharynx. This is a region we have already become acquainted with in dealing with younger embryos, but its relations are so important that some review of them here will not be amiss. On either side the pharyngeal lumen shows a series of extensions or bays known as the *pharyngeal pouches* (Fig. 90). Each pharyngeal pouch lies opposite an external gill furrow or *branchial groove* (Figs. 95 and 97, B). This leaves in these areas only a thin layer of tissue separating the pharyngeal lumen from the outside. This layer is composed internally of pharyngeal entoderm and externally of the ectoderm of the bottom of the branchial groove (Fig. 97, B). There may or may not be a small amount of mesenchyme between these two epithelial layers of the *gill plate*. Sometimes the more cephalic gill plates actually break through, establishing transitory open gill clefts.

Between adjacent gill grooves or clefts the lateral walls are greatly thickened and filled with closely packed mesenchymal cells (Fig. 96, B). These thickened areas are known as the *branchial arches* (also sometimes called visceral arches or gill arches). One of the most important relationships to grasp in the study of young embryos is the manner in which the *aortic arches* lie embedded in the tissue of the branchial arch of corresponding number. The stereogram appearing in Fig. 95 was planned especially to emphasize these relationships. It should be intensively studied in conjunction with the section diagrammed below it until a clear three-dimensional concept of the relations in this region has been grasped.

The Pharyngeal Derivatives   Several structures, which do not become parts of the digestive system, arise in the pharyngeal region. Nevertheless the origin of their epithelial portions from foregut entoderm and their early association with this part of the gut tract make it convenient to consider them in connection with the digestive system.

The *thyroid gland* arises from the floor of the pharynx as a median diverticulum which makes its appearance at a cephalo-caudal level between the first and second pair of pharyngeal pouches (Fig. 90). Toward the end of the fourth day the thyroid evagination has become saccular (Fig. 89). Later

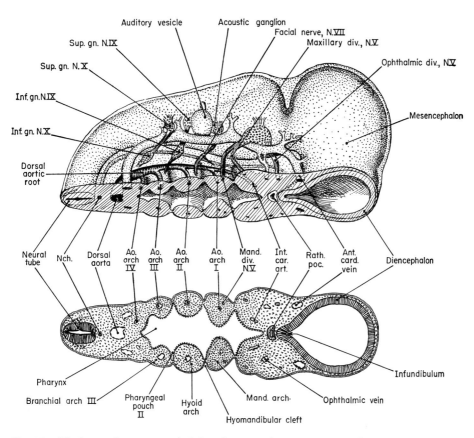

Fig. 95.   The lower drawing is a slightly schematized representation of a transverse section through the pharyngeal region of a 3½-day chick. This section is keyed to a stereogram showing cephalic structures above the level of the section. The illustration was planned primarily to show the relations of the aortic arches to the branchial arches.

    Abbreviations: Int. car. art., internal carotid artery; *Mand. div. N. V,* mandibular division of fifth (trigeminal) cranial nerve; *Nch.,* notochord; *Rath. poc.,* Rathke's pocket. *(Method of presentation suggested by some of the drawings in Huettner's "Fundamentals of Comparative Embryology of the Vertebrates," The Macmillan Company, New York.)*

in development it moves caudad in the body, but for a time retains its connection with the root of the tongue by a narrowed portion known as the *thyroglossal duct.* The paired evaginations which arise posteriorly from the fourth pharyngeal pouches are known as *postbranchial bodies.* Their significance is not altogether clear. Homologous evaginations in mammals have been said by some investigators to contribute to the formation of the thyroid gland. In the chick the postbranchial bodies do not appear to form thyroid tissue. Moreover, recent work on these structures in mammals indicates that in that group also, they probably do not give rise to thyroid tissue as formerly believed, in spite of the very close positional relation which comes to exist between them and the median thyroid evagination.

The two pairs of *parathyroid glands*, also, arise as bud-like outgrowths from the pharynx. One pair of parathyroids is budded off the caudal faces of the third pharyngeal pouches, and the other pair arises in a similar manner from the fourth pouches. These parathyroid primordia are, however, not usually recognizable until later stages of development than those here under consideration.

The *thymus* of the chick is barely indicated, if present at all, on the fourth day of incubation. It takes its origin primarily from diverticula arising from the posterior faces of the third and fourth pharyngeal pouches. The original epithelial character of the thymus is soon largely lost in an extensive ingrowth of mesenchyme, and the organ becomes chiefly lymphoid in its histologic characteristics.

The Trachea   The first indication of the formation of the respiratory system appears in 3-day chicks as a midventral groove in the pharynx. Beginning just posterior to the level of the fourth pharyngeal pouches and extending caudad (Fig. 96, *B*), this *laryngo-tracheal groove* deepens rapidly and by closure of its dorsal margins becomes separated from the pharynx except at its cephalic (laryngeal) end. The tube thus formed is the *trachea* (Figs. 89 and 90), and the opening which persists between the laryngeal end of the trachea and the pharynx is the *glottis*. The original entodermal evagination gives rise only to the epithelial lining of the trachea, the supporting structures of the tracheal walls being derived from the surrounding mesenchyme.

The Lung-buds   The tracheal evagination grows caudad and bifurcates to form a pair of lung-buds. As the lung-buds develop they grow into the loose mesenchyme on either side of the midline. This relationship of mesenchyme to growing entodermal epithelium, as seen here in the lungs, and also in the developing gastrointestinal tract, is an interesting one. Apparently the formation of tubular structures, such as the bronchi of the lungs or the ducts and secreting tubules of digestive glands, is dependent on the presence of mesoderm about the growing entoderm. If entoderm is grown in a culture without the presence of any mesoderm, it tends to spread out as a flat, unorganized sheet of sprawling cells. If, however, mesenchymal cells are added to such cultures, the entodermal epithelium forms tubular structures. These tubular structures, moreover, will take on the characteristics of the lung, or a digestive gland, or the epithelial lining of the gut tract according to the primordial region from which the original entodermal cells were taken. Thus again, this time, at the level of tissue differentiation, we see the influence of one growing part of the embryo on another.

In the caudo-lateral growth of the primordial lung-buds the surrounding splanchnic mesoderm is pushed ahead of them and comes to constitute their outer investment (Fig. 97, D). The entodermal buds give rise only to the epithelial lining of the bronchi and the air passages and air chambers of the lungs. The connective tissue stroma of the lungs is derived from mesenchyme immediately surrounding the entodermal buds, and their pleural covering is formed from their primary investment of splanchnic mesoderm (Fig. 97, D).

The Esophagus and Stomach   Immediately caudal to the glottis is a narrowed region of the foregut, which becomes the esophagus, and farther caudally a slightly dilated region, which becomes the stomach (Fig. 90). The concentration of mesenchymal cells about the entoderm of the esophageal and gastric regions foreshadows the formation of their muscular and connective tissue coats (Fig. 97, C, E).

The Liver   In all vertebrates the liver arises as a diverticulum from the ventral wall of the gut, a little caudal to the stomach. In chick embryos the hepatic primordium can first be recognized at about the 22-somite stage. It appears just as the part of the gut from which it arises is acquiring a floor by the concrescence of the margins of the anterior intestinal portal. As a result the evagination which is the primordium of the liver is located for a short time on the lip of the intestinal portal and grows cephalad toward the confluence of the omphalomesenteric veins just before they enter the sinus venosus. As closure of the gut floor is completed the hepatic diverticulum comes to lie in its characteristic position in the ventral wall of the gut. In embryos of 4 days the original evagination has grown out in the form of branching cords of cells and become quite extensive in mass (Fig. 90). In its growth the liver pushes ahead of itself the splanchnic mesoderm which surrounds the gut, with the result that the liver from its first appearance is invested by mesoderm (Figs. 89, 96, C, and 97, E).

The proximal portion of the original evagination remains open to the intestine, and serves as the duct of the liver. This primitive duct later undergoes regional differentiation and gives rise in the adult to the *common bile duct*, to the *hepatic and cystic ducts*, and to the *gallbladder*. The cellular cords which bud off from the diverticulum become the secretory units of the liver *(hepatic tubules)*.

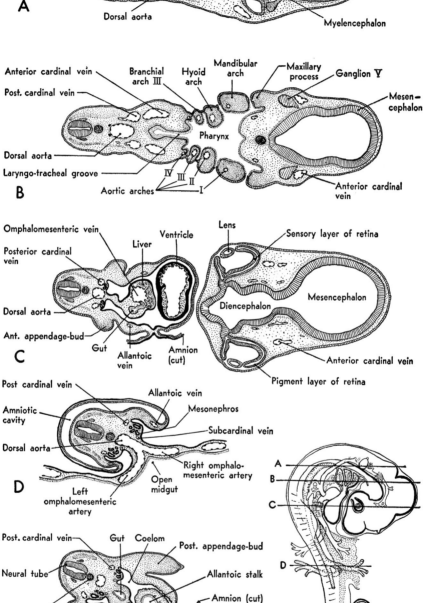

Fig. 96. Diagrams of five representative transverse sections of a 3-day chick. The locations of the sections are indicated on the small outline sketch of the entire embryo.

The same process of concrescence which closes the floor of the foregut involves the proximal portion of the omphalomesenteric veins which, when they first appear, lie in the lateral folds of the anterior intestinal portal (Fig. 75). As the intestinal portal moves caudad in the lengthening of the foregut, the proximal portions of the omphalomesenteric veins are brought together in the midline and become fused. The fusion extends caudad nearly to the level of the yolk-stalk (Fig. 101). Distal to this point they retain their original paired condition. In its growth the liver surrounds the fused portion of the omphalomesenteric veins (Figs. 91, 96, *C*, and 97, *E*). This early association of the omphalomesenteric veins with the liver foreshadows the way in which blood channels coming in from the vitelline vascular plexus are to be involved in the establishment of the hepatic-portal circulation of the adult.

The Pancreas   The pancreas is derived from evaginations appearing in the walls of the intestine at the same level as the liver diverticulum. There are three pancreatic buds, a single median dorsal, and a pair of ventro-lateral buds. The dorsal evagination appears at about 72 hours, the ventro-lateral evaginations toward the end of the fourth day. The dorsal pancreatic bud arises from gut entoderm directly opposite the liver diverticulum (Fig. 90) and grows into the dorsal mesentery (Fig. 112, *G*, and 113). The ventro-lateral buds arise close to the point where the duct of the liver connects with the intestine so that the ducts of the liver and the ventral pancreatic ducts open into the intestine by a common duct *(ductus choledochus).* Later in development the masses of cellular cords derived from the three pancreatic primordia grow together and become fused into a single glandular mass, but, in birds, usually two and in rare cases all three of the original ducts persist in the adult.

The Midgut Region   In chicks of four days the enteric tract shows no local differentiation from the level of the liver to the cloaca except where the yolk-sac is attached. All of the gut tract between the stomach and the yolk-stalk, and the anterior fourth of the gut which lies caudal to the yolk-stalk, are destined to become the small intestine. The posterior two-thirds of the hindgut becomes large intestine and cloaca.

The Cloaca   The beginning of the formation of the cloaca is indicated in chicks of four days incubation by a dilation of the posterior portion of the hindgut (Fig. 90). Although extensive differentiation in the cloacal region does not appear until later in development, certain of its fundamental relationships are established at this stage.

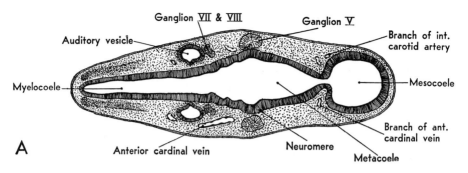

A

Ganglion VII & VIII
Auditory vesicle
Ganglion V
Branch of int. carotid artery
Myelocoele
Mesocoele
Anterior cardinal vein
Neuromere
Branch of ant. cardinal vein
Metacoele

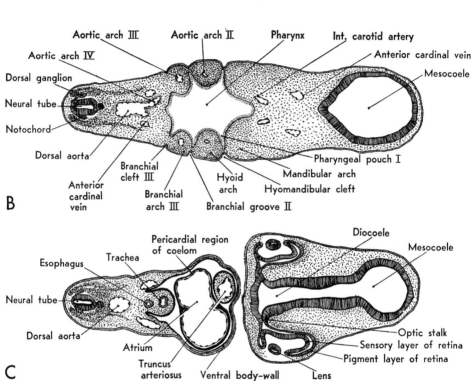

B

Aortic arch III
Aortic arch II
Pharynx
Int. carotid artery
Aortic arch IV
Anterior cardinal vein
Dorsal ganglion
Mesocoele
Neural tube
Notochord
Dorsal aorta
Pharyngeal pouch I
Branchial cleft III
Hyoid arch
Mandibular arch
Anterior cardinal vein
Branchial arch III
Hyomandibular cleft
Branchial groove II

C

Pericardial region of coelom
Diocoele
Trachea
Esophagus
Mesocoele
Neural tube
Dorsal aorta
Optic stalk
Atrium
Sensory layer of retina
Truncus arteriosus
Pigment layer of retina
Ventral body-wall
Lens

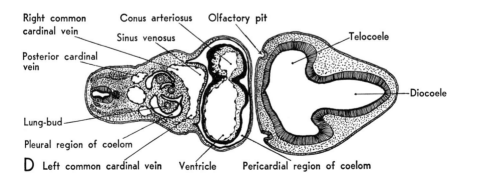

D

Right common cardinal vein
Conus arteriosus
Olfactory pit
Sinus venosus
Telocoele
Posterior cardinal vein
Lung-bud
Diocoele
Pleural region of coelom
Left common cardinal vein
Ventricle
Pericardial region of coelom

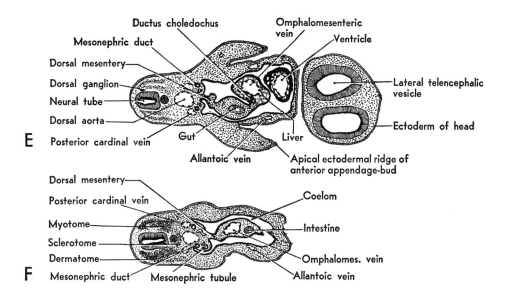

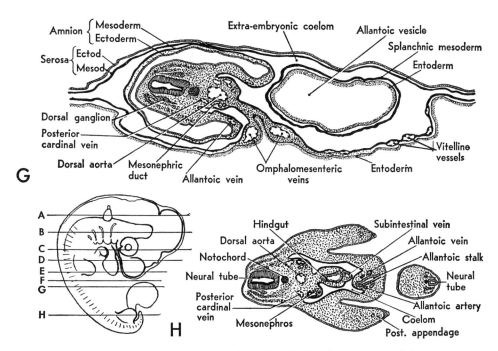

Fig. 97.   Diagrams of transverse sections of a 4-day chick. The locations of the sections are indicated on a small outline sketch of the entire embryo.

The cloaca of an adult bird is the common chamber into which the intestinal contents, the urine, and the products of the reproductive organs are received for discharge. The first appearance of the cloaca in the embryo as a dilated terminal portion of the gut establishes at the outset the relations of cloaca and intestine that persist in the adult.

Although the urinary system is not at this stage developed to conditions which resemble those in the adult, the parts of it which have been established are already definitely associated with the cloaca. The proximal portion of the allantoic stalk, which is the homolog of the urinary bladder of mammals, opens directly into the cloaca (Fig. 90). When the urinary system of the embryo is considered, we shall see that the ducts which drain the developing excretory organs also open into the cloacal region on either side of the allantoic stalk.

There is at this stage but little indication of the formation of the gonads. The relation of the sexual ducts to the cloaca can be made out only by the study of older embryos.

The Proctodaeum and the Cloacal Membrane   Indications of the formation of the cloacal opening to the outside appear during the fourth day of incubation. Its establishment is accomplished in much the same manner as the establishment of the oral opening. A ventral outpocketing of the hindgut arises just caudal to the point at which the allantoic stalk opens into the cloaca (Fig. 90). At the same time a depression appears in the overlying ectoderm. The external depression which grows in toward the gut pocket is known as the *proctodaeum*. The double epithelial layer formed by the meeting of gut entoderm with proctodaeal ectoderm is the cloacal membrane. The formation of the proctodaeum and the cloacal membrane clearly indicates the location of the future cloacal opening although an open communication is not established by the rupture of the cloacal membrane until considerably later. The cloacal opening does not form at the extreme posterior end of the hindgut and there is, therefore, a post-anal pocket of the hindgut suggestive of the pre-oral pocket of the foregut (Fig. 89).

V.   THE CIRCULATORY SYSTEM

Interpretation of the Embryonic Circulation   The embryonic circulation is difficult to understand only when the meaning of its arrangement is overlooked. If one bears in mind the significance of the circulatory system in organic economics and the fact that any embryo must go through certain ancestral phases of organization before it can arrive at its adult structure, the changes in the arrangement of vascular channels during the course of development form a coherent and logical story.

In the embryo, as in the adult, the main vascular channels lead to and from the centers of metabolic activity. The circulating blood carries food from the organs concerned with its absorption to parts of the body remote from the source of supplies. It also carries oxygen to all the tissues of the body, from organs which are especially adapted to facilitate the taking of oxygen into the blood; and waste materials, from the places of their liberation, to the organs through which they are eliminated. One of the primary reasons the arrangement of the vessesl in an embryonic bird or mammal differs so much from that in the adult is the fact that the embryo lives under conditions totally unlike those which surround its parents. Its centers of metabolic activity are, therefore, different; and, since the course of its main blood vessels is determined by these centers, the vascular plan is different. No such profound changes as we find in birds and mammals occur between the embryonic and the adult stages in the circulation of a fish where embryo and adult are both living under similar environmental conditions.

The organs which in the adult carry out such functions as digestion and absorption, respiration, and excretion are extremely complex and highly differentiated. They are for this reason slow to attain their definitive condition and are not ready to become functional until toward the close of the embryonic period. Moreover the conditions which surround certain of the developing organs during embryonic life would prevent their becoming functional even were they sufficiently developed so to do. Suppose the lungs, for example, were functionally competent at an early stage of development. The fact that the embryo is reliving ancestral conditions submerged in the amniotic fluid renders its lungs as incapable of functioning as those of a man under water. Likewise the developing stomach of a chick encased in its shell can receive no raw food materials by way of the mouth. Further examples are not necessary to make it obvious that, were the embryo dependent on the same organs which carry on metabolism in the adult, its development would be at an impasse.

Nevertheless an embryo must solve the problem of existence during the time in which it is building up a set of organs similar to those of its parents. The chick must have not only the raw food material supplied by its mother in the form of yolk, it must have in addition a means of digesting the yolk, absorbing it, and carrying it to the places where it can be utilized. Furthermore the utilization of food material to produce the energy expressed in growth depends on the presence of oxygen. So for growth there must be a means of securing oxygen and carrying it, as well as food, to all parts of the body. Nor can continued growth go on unless the waste products liberated by the growing tissues are eliminated. At the outset of its development the embryo must, therefore, establish organs for the digestion and absorption of food, the securing of oxygen, and the elimination of waste products. In various types of embryos these functions are carried out by the yolk-sac or the allantois or by both together, in a manner depending on the exigencies of the embryo's living conditions. These structures are relatively simple in comparison with the organs which carry out the corresponding fuctions in the adult, but their activities are so essential and so extensive that the vascular channels supplying them are a dominating feature in the embryonic circulation.

While main circulatory channels are thus always established in relation to centers of metabolic activity, we find in many situations in the embryo an unmistakable phylogenetic impress on the manner of their establishment or on the details of their arrangement. Take, for example, the aortic arches. The blood leaving the ventrally located heart must pass around the gut to reach the dorsally located aorta. In fishes six pairs of aortic arches encircle the pharynx, breaking up in the gills into capillaries which carry out the indispensable function of oxygenating the blood. Once an animal has replaced its gills with lungs, it makes little difference functionally whether or not the blood passes by each gill cleft on its way from heart to aorta. In adult birds and mammals we find this communication simplified to a single main aortic arch. But in the embryos of both birds and mammals the whole series of symmetrical aortic arches appear for a time, encircling the pharynx and passing in close relation to vestigial gill clefts. This can be interpreted only as a recapitulation of ancestral conditions—conditions which, although they have ceased to be of functional importance, appear nevertheless as a developmental phase on the way to a more highly differentiated plan of adult structure. Our interpretation of the aortic arches, then, would take into consideration first the fundamental functional necessity of a channel connecting the ventrally located heart with the dorsally located aorta. Second, looking at the striking arrangement of the series of aortic arches in relation to the gill arches and clefts, one sees a repetition of the structural plan which existed in water-living ancestral forms. Thus in the end we are led back again to functional significance, for the relationship of the aortic arches to the gill arches writes into the story of individual development an unequivocal record of the evolutionary phase when the gills were a center of primary metabolic importance.

Applied to the development of the circulatory system as a whole, this tendency to repeat phyletic history means that the earliest form in which the circulatory mechanism of a bird or a mammal appears is patterned after a more primitive type and, therefore, cannot be a miniature of adult conditions. The simple tubular heart pumping blood out over aortic arches to be distributed over the body by the aorta and returned to the posterior part of the heart by a bilaterally symmetrical venous system, in short the foundational vascular plan which we see in sauropsidan embryos, is essentially the plan of the circulation in fishes. When we realize this, we are puzzled neither by the early appearance of a full complement of aortic arches nor by their subsequent disappearance to make way for a new respiratory circulation in the lungs. Starting from a logical beginning in simpler ancestral conditions, we see the march of progress toward the consummation of embryonic life with the attainment of an organization like that of the immediate parent. And at each turning point the direction of further advance is sharply limited. Development can proceed only along lines that permit the maintenance of an efficient metabolism, both in the temporary organs characteristic of embryonic life and in the slowly developing organs which must be prepared to meet adult living conditions.

Finally, we must bear in mind that only relatively late in development, as the various organs become established in their adult relationships, can we expect to see the emergence of the vascular plan characteristic of the adult. Following the primitive embryonic stage there must be transitional phases when functioning embryonic channels are present side by side with channels being prepared to take over their activity in the adult. For all changes in the circulatory system of a living organism must be gradual. Any changes which were sufficiently abrupt to interfere with the circulation would result in disaster for the embryo. Even slight curtailment of the normal blood supply to any region would cause its growth to cease; any marked local decrease in the circulation would result in local atrophy or malformation; complete interruption of any important circulatory channel, even for a short time, would inevitably mean the death of the embryo.

Main Routes of the Embryonic Circulation   Before taking up the various parts of the circulatory system in detail it might be well to consider briefly the main routes of the embryonic circulation. The circulation of young chick embryos involves three main arcs of which the heart is the common center and pumping station. One of these circulatory arcs, the vitelline, carries blood to the yolk-sac where food materials are absorbed and then returns the food-laden blood to the heart for distribution within the embryo. Another arc carries blood to and from the allantois. The distal portion of the allantois lies close beneath the egg shell, and the blood circulating in the allantoic vessels is thereby brought into a location where interchange of gases can be carried on with the air which penetrates the shell (Fig. 98). It is in the allantoic circulation that the blood gives off its carbon dioxide and acquires a fresh supply of oxygen. The allantoic circulation is also the embryo's means of eliminating nitrogenous waste material from the blood. The remaining circulatory arc is confined to the body of the embryo. The intra-embryonic circulation has many distributing and collecting vessels, but all of them are alike in function in that they bring food material and oxygen to, and carry waste material from, the various parts of the developing body. Nowhere in their course are the vessels of the intra-embryonic circulation involved in adding food material or oxygen to that already contained in the blood they convey, and nowhere do they free the blood from waste materials until well along in development, when the nephroi become functional.

In the heart the blood from the three circulatory arcs is mingled. As it leaves the heart the mixed blood is not as rich in food material as the blood coming in through the omphalomesenteric veins, nor as free from waste materials and as rich in oxygen as the blood returned over the allantoic veins.[1] Its condition of serviceability to the embryo is, however, constantly maintained at a good average by the incoming vitelline and allantoic blood.

The Vitelline Circulation   The earliest indication of blood and blood vessel formation is at the chick's source of food supply. Blood islands appear in the extra-embryonic splanchnopleure of the yolk-sac toward the end of the first day of incubation and rapidly become differentiated to form vascular endothelium enclosing central clusters of primitive blood corpuscles (Fig. 58). By extension and anastomosing of neighboring islands, a plexus of blood channels is formed in the yolk-sac. Further extension of the vitelline

---

[1] There is a tendency among students who have done but little work on the circulatory system to regard any vessel which carries oxygenated blood as an artery, and any vessel which carries blood poor in oxygen and high in carbon dioxide content as a vein. This is not entirely correct even for the circulation of adult mammals on which the conception is based. In comparative anatomy and especially in embryology it is far from being the case. It is necessary, therefore, in dealing with the circulation of the embryo to eradicate this not uncommon misconception.

The differentiation between arteries and veins which holds good for all forms, both embryonic and adult, is based on the structure of their walls, and on the direction of their blood flow with reference to the heart. An artery is a vessel carrying blood away from the heart under a relatively high, fluctuating pressure due to the pumping of the heart. Correlated with the pressure conditions in it, its walls are heavily reinforced by elastic tissue and smooth muscle. A vein is a vessel carrying blood toward the heart under a relatively low and constant pressure from the blood welling into it from capillaries. Correlated with such pressure conditions, the walls of a vein have much less smooth muscle than artery walls, and their connective tissue is predominantly of the nonelastic type.

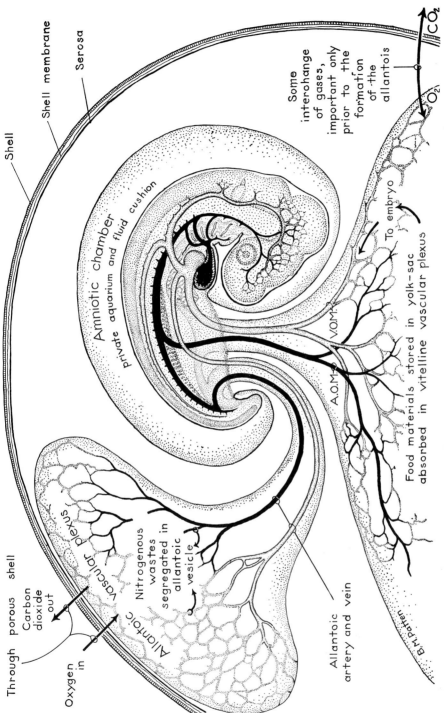

Fig. 98. Schematic diagram showing arrangement of main circulatory channels in a young chick embryo. The sites of some of the extra-embryonic interchanges important in its bio-economics are indicated by the labeling. The vessels within the embryo carry food and oxygen to all its growing tissues and relieve them of the waste products incident to their metabolism. [*From Patten, National Sigma Xi Lecture, reprinted in The Am. Scientist,* **39** (1951).]

Shell

Shell membrane

Serosa

Amniotic chamber
private aquarium and fluid cushion

Some interchange
of gases,
important only
prior to the
formation
of the
allantois

CO₂

O₂

To embryo

V.O.M.

A.O.M.

Food materials stored in yolk-sac
absorbed in vitelline vascular plexus

B.M.Patten

Through porous shell
Carbon dioxide out

Oxygen in

Allantoic vascular plexus

Nitrogenous wastes
segregated in
allantoic
vesicle

Allantoic
artery and vein

plexus brings it into communication with the omphalomesenteric veins which have been developed in the embryo as caudal extensions of the endocardial primordia (Figs. 55 and 60).

Toward the end of the second day of development the omphalomesenteric arteries establish communication between the dorsal aortae and the vitelline plexus. (See Chap. 9 and Figs. 65 and 67.) There is now a system of open channels leading from the embryo to the yolk-sac, and back again to the embryo. With the completion of these channels the heart, which has for some hours been building up the efficiency of its pulsation, is able to set the blood in circulation. It is thus, as we have seen, at about the 40th hour of incubation that the blood cells formed in the yolk-sac are for the first time carried into the body of the embryo.

The course of the vitelline circulation in chicks of 4 days is shown diagrammatically in Fig. 99. Circulating in the rich plexus of small vessels on the yolk, the blood finally makes its way either directly into one of the larger vitelline veins, or into the sinus terminalis which acts as a collecting channel, and then through the sinus terminalis to one of the vitelline veins. The vitelline veins converge toward the yolk-stalk where they empty into the omphalomesenteric veins. The omphalomesenteric veins, at first paired throughout their entire length, have been brought together proximally by the closure of the ventral body-wall and become fused to form a median vessel within the body of the embryo. It is through this vessel that the blood returning from the vitelline circuit eventually reaches the heart. In the heart the blood of the vitelline, intra-embryonic, and allantoic circulations is mingled. The mixed blood passes out by the ventral aorta and the aortic arches into the dorsal aorta. Leaving the dorsal aorta through the vitelline arteries a certain portion of this blood is returned to the yolk-sac for another circuit through it.

It should not be inferred that the blood stream "picks up" deutoplasmic granules and carries them to the embryo. The acquisition of food material by the blood depends on the activities of the entodermal cells lining the yolk-sac. These cells secrete digestive enzymes which break down the deutoplasmic granules. The liquefied material is then absorbed by the yolk-sac cells and transferred to the blood (Fig. 100). The blood carries the food material in soluble form to the embryo where it is finally assimilated.

The Allantoic Circulation   The allantoic arteries arise by the prolongation and enlargement of a pair of vessels arising ventrally from the aorta at the level of the allantoic stalk. Their size increases rapidly as the allantois

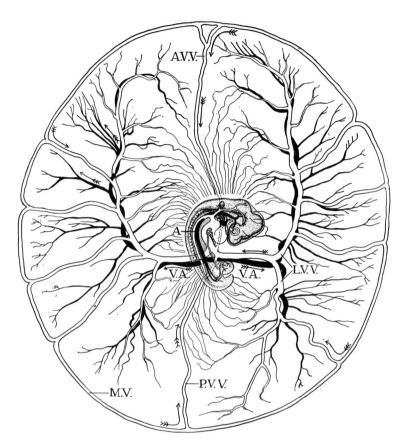

Fig. 99.   Diagram to show course of vitelline circulation in chick of about 4 days. For the intra-embryonic vessels, see Fig. 101.

Abbreviations: A, dorsal aorta; *A. V. V.*, anterior vitelline vein; *L. V. V.*, lateral vitelline vein; *M. V.*, marginal vein (sinus terminalis); *P. V. V.*, posterior vitelline vein; *V. A.*, vitelline artery. The direction of blood flow is indicated by arrows. (After Lillie.)

increases in extent. From them the blood is distributed in a rich plexus of vessels which ramify in the mesoderm of the allantois (Figs. 73, *C*, and 98).

The situation of the allantois directly beneath the porous shell is such that the blood can carry on interchange of gases with the outside air. It is in the rich plexus of small allantoic vessels where the surface exposure is very great that the blood gives off its carbon dioxide and takes up oxygen (Fig. 98).

At a later stage of development the ducts of the embryonic excretory

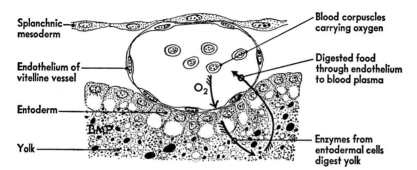

Fig. 100.   Semischematic high-power drawing of a yolk-sac blood vessel and the adjacent entoderm and yolk. The labeling and the arrows indicate the important processes going on in such an area.

organs open into the allantoic stalk near its cloacal end. As the excretory organs become functional, the allantoic vesicle becomes the repository for the nitrogenous waste materials eliminated through them. The watery portion of the waste materials is passed off by evaporation. The remaining solids are deposited in the allantoic vesicle. They accumulate in the extra-embryonic portion of the allantois and there remain until that portion of the allantois is discarded at the close of embryonic life.

The blood from the allantois is collected and returned to the heart over the allantoic veins. From the distal portion of the allantois the smaller veins converge and unite into two main vessels, right and left, which enter the body of the embryo with the allantoic stalk (Fig. 97, *H*). After their entrance into the body the allantoic veins extend cephalad in the lateral body-walls (Figs. 97, *G–E*, and 101). They enter the sinus venosus on either side of the entrance of the omphalomesenteric vein.

The Intra-embryonic Circulation   The earliest vessels of the intra-embryonic circulation to appear are the large vessels communicating with the heart. In chicks of 33 hours the ventral aorta leads off from the heart cephalically and bifurcates ventral to the pharynx giving rise to a single pair of aortic arches. These first aortic arches pass dorsad around the antero-lateral walls of the pharynx and are continued caudally along the dorsal wall of the gut as the paired dorsal aortae (Figs. 57, 64, and 79).

When, toward the end of the second day of incubation, branchial clefts and branchial arches appear, the original pair of aortic arches comes to lie in the mandibular arch. In each of the branchial arches posterior to the mandibular, new aortic arches will be formed connecting the ventral aortae with the dorsal aortae. By 50 hours two pairs of aortic arches are present and a third is beginning to form (Figs. 75 and 80). In chicks of 60 hours' incubation three aortic arches have been completed, and the fourth is usually starting to form (Fig. 78). By the end of the fourth day a rudimentary fifth and a well-developed sixth arch are present (Fig. 101). From their first appearance the fifth aortic arches are very small and they soon disappear altogether. The first and second pairs of aortic arches have by this time suffered a marked diminution in size, which is indicative of their final disappearance. In many embryos of this age range the first arches, and in a few the second also, have disappeared altogether (Fig. 102). This leaves only the third, fourth, and sixth pairs of aortic arches. These arches persist intact for some time, and parts of them remain permanently, being incorporated in the formation of the aortic arch and the main vessels arising from it, and in the roots of the pulmonary arteries.

In embryos late in the second day we saw the appearance of plexiform channels extending from the first aortic arch toward the forebrain (Figs. 64 and 79) and noted that they foreshadowed the formation of the *internal carotid arteries*. By 60 hours the internal carotid artery is established as a definite trunk extending toward the brain from the point where the first aortic arch turns into the dorsal aortic root (Fig. 78). With the regression of the first two aortic arches during the fourth day of development, the dorsal aortic root at this level is appropriated as a feeder to the original internal carotid artery, thereby making it appear to originate where the third aortic arches merge with the dorsal aortic roots (Fig. 101). The meshwork of small vessels fed by the internal carotid artery and developing in close relation to the brain is one of extraordinary richness (Fig. 102). In embryos of this age it is possible to see suggestions of certain of the main named branches characteristic of the adult, but for the most part they are represented only by primordial capillary plexuses.

The regression of the first two aortic arches plays a still more prominent role in the formation of the *external carotid arteries*. When these two arches lose connection with the dorsal aortic roots, the part of the ventral aortic roots which formerly fed them persists (Fig. 101). These vessels represent the main stems of the external carotid arteries which later develop many important branches supplying the region of the mandible and the front of the neck and the face.

In reptiles, birds, and mammals the main adult vessels which connect the heart with the dorsal aorta are derived from the fourth pair of aortic arches of the embryo. The paired condition of these arches persists as the adult condition in reptiles, but in birds and mammals one of the arches de-

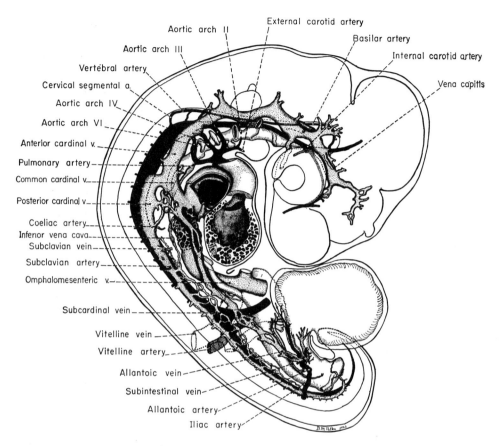

Fig. 101. Reconstruction of circulatory system of 4-day chick (original ×51, reproduced ×18). From the same embryo as that represented in Figs. 90 and 91. These three illustrations should be studied together.

generates before the end of embryonic life. In birds the left arch degenerates leaving the right one as the arch of the adult aorta; in mammals the right arch degenerates leaving the left as the aortic arch of the adult aorta.

The dorsal aortae, at first paired, later become fused to form a median vessel. As we have seen, the fusion begins at cardiac level. It extends cephalad, but a short distance, never involving the region of the dorsal aortic roots where they receive the aortic arches. Caudally the aortae eventually become fused throughout their entire length.

Early in development the aorta gives rise to a segmentally arranged series of small vessels which extend into the dorsal body-wall (Fig. 102). At the level of the anterior appendage-buds several pairs of these *dorsal*

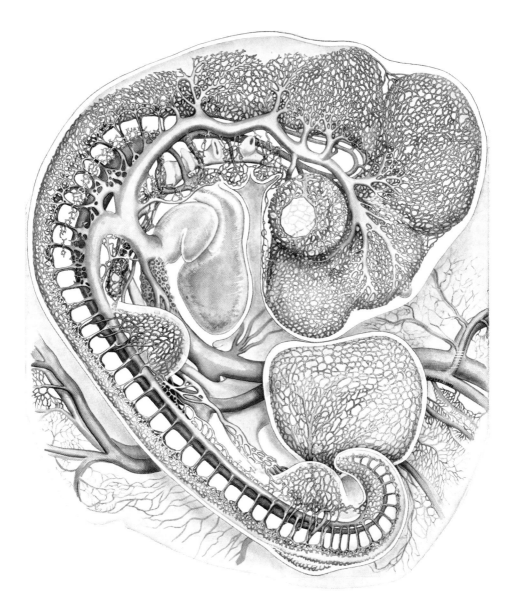

Fig. 102. The blood vessels of a 4-day chick. The basis of the illustration was the same wax-plate reconstruction from which Fig. 91 was drawn. The smaller vessels and the primordial capillary plexuses were added from injected specimens. The labeled diagram of Fig. 101 will serve as a means of identifying the main vessels.

segmental arteries extend into the wing buds (Fig. 102). Later one of these becomes enlarged to form the *subclavian arteries*.

Coincident with the development of the allantois, a pair of *ventral segmental vessels* opposite the allantoic stalk become enlarged and extend into it as the *allantoic arteries*. These, as we have seen, feed a rich plexus of small vessels which, with the growth of the allantois, come to lie close beneath the porous shell and become concerned with gaseous interchanges (Fig. 98).

The blood supply to the posterior appendage starts as a plexus fed by several segmental vessels at the level of the appendage-bud (Fig. 102). Later this primordial plexus comes to be fed by a branch arising from the allantoic artery. This new vessel becomes the *external iliac artery*.

In the adult, three vessels arising from the dorsal aorta supply the abdominal viscera. They are the *coeliac, superior mesenteric,* and *inferior mesenteric arteries*. In 4-day chicks these vessels are usually represented only by the omphalomesenteric arteries. The omphalomesenteric arteries arise as paired vessels (Fig. 75), but in the closure of the ventral body-wall of the embryo they are brought together and fused to form a single vessel which runs in the mesentery from the aorta to the yolk-stalk (Fig. 96, *D*). With the atrophy of the yolk-sac the proximal part of the omphalomesenteric artery persists as the superior mesenteric of the adult. The coeliac and the inferior mesenteric arteries arise from the aorta independently, usually at a somewhat later stage. Occasionally, however, the coeliac can be identified in 4-day embryos (Fig. 101).

As development progresses, any arterial trunk growing into an organ will exhibit a pattern of branching which bears a definite relation to the structure of the organ it is supplying. The vascular sprouts tend to follow along the developing connective tissue framework of the organ, assuming a configuration dictated by its arrangement. This can very prettily be demonstrated experimentally by the transplantation of an organ primordium before it has become vascularized. When this is done, the host vessels which grow in and supply the transplanted organ will be found to show the same basic arrangement that would have appeared if the organ had been allowed to develop undisturbed.

The *anterior* and *posterior cardinal veins* are the principal afferent systemic vessels of the early embryo. They appear toward the end of the second day as paired vessels extending cephalically and caudally on either side of the midline. At the level of the heart the anterior and posterior cardinal veins of the same side of the body become confluent in the common cardinal veins and turn ventrad to enter the sinus venosus (Figs. 64 and 78).

Chicks of 4 days show little change in the relationships of the cardinal veins (Figs. 101 and 102). Later in development the proximal ends of the anterior cardinals become connected by the formation of a new transverse vessel and empty together into the right atrium of the heart. Their distal portions remain in the adult as the principal afferent vessels (*internal jugular veins*) of the cephalic region.

The posterior cardinals lie at first in the angle between the somites and the lateral mesoderm (Fig. 76, *D*, *E*). When the mesonephroi develop from the intermediate mesoderm, the posterior cardinal veins lie just dorsal to them throughout their length (Figs. 96, *D*, *E*, 97, *E–H*, and 110, *C*). Situated ventrally in the mesonephroi are small irregular vessels roughly paralleling the posterior cardinals and anastomosing freely with them. These are the subcardinal veins (Fig. 101). They are a relic of the renal portal circulation of more primitive ancestral forms, and are of importance in the embryos of birds and mammals chiefly because of the way they are involved in the formation of the posterior vena cava. In young embryos the posterior cardinals are the main afferent vessels of the posterior part of the body. Later in development they are replaced in this function by the posterior vena cava. In 4-day chicks only the upper part of the posterior vena cava is indicated. It appears as a slender vessel extending from the right subcardinal vein through a fold in the base of the dorsal mesentery to anastomose with venous channels in the liver. The formation of the vena cava, in part as a new vessel, and in part by appropriation of already existing vessels, and the changes by which the posterior cardinals become reduced and broken up to form small vessels with new associations belong to stages of development beyond the scope of this book.

The Heart   The heart in adult vertebrates is a ventral, upaired structure. Its origin in the chick from paired primordia is correlated with the way the young embryo lies spread out on the yolk surface. When the ventral body-wall is completed by the folding together of layers which formerly extended to right and left over the yolk, the paired primordia of the heart are brought together in the midline. Their sequential fusion establishes the heart as an unpaired structure lying in the characteristic ventral position. (See Chap. 8 and Figs. 59, 60, and 61.)

After the fusion of its paired primordia the heart is a nearly straight, double-walled tube (Figs. 53 and 103, *A*). The primordial endocardium of the heart has the same structure and arises in the same manner as the endothelial walls of the primitive embryonic blood vessels with which it is directly continuous. The epi-myocardial layer of the heart is an outer investment which surrounds and reinforces the endocardial wall. As development progresses, the epi-myocardium becomes greatly thickened and is finally differentiated into two layers, a heavy muscular layer, the myocardium, and a thin non-muscular covering layer, the epicardium.

In the apposition of the paired primordia of the heart to each other, the splanchnic mesoderm from either side of the body comes together dorsal and ventral to the heart. The double-layered supporting membranes thus formed are known as the dorsal mesocardium and the ventral mesocardium respectively (Fig. 59, D). The ventral mesocardium disappears shortly after its formation, leaving the heart suspended in the body cavity by the dorsal mesocardium (Fig. 59, E). Somewhat later the dorsal mesocardium also disappears except at the caudal end of the heart. Thus the heart comes to lie in the pericardial cavity unattached except at its two ends. The cephalic end of the heart remains fixed with reference to the body of the embryo where the ventral aorta lies embedded ventral to the floor of the pharynx, and the caudal end of the heart is fixed by the persistent portion of the dorsal mesocardium and the omphalomesenteric veins.

The straight tubular condition of the heart persists but a short time. As the heart is lengthened, the unattached ventricular region becomes dilated and is bent out of the midline toward the embryo's right while the fixed outlet of the truncus arteriosus and the firmly anchored sino-atrial end are held in their original median position (Fig. 104, A–E). This bending of the heart to form a U-shaped tube begins to be apparent in embryos of 30 hours and becomes rapidly more conspicuous, until by 40 hours the ventricular region of the heart lies well to the right of the embryo's body (Fig. 103, B–E).

The bending of the heart to the side involves a considerable factor of "mechanical expediency." The initiation of the bending process depends on the fact that the heart is becoming elongated more rapidly than is the chamber in which it lies fixed by its two ends. The fact that the bending takes place to the side rather than dorsally or ventrally may be attributed to the impediment offered to its dorsal bending by the body of the embryo and to its ventral bending by the yolk.

The lateral bending of the heart attains its greatest extent at about 40 hours of incubation. At this stage torsion of the body of the embryo begins to change the mechanical limitations in the cardiac region. As the embryo comes to lie on its left side, the heart is no longer pressed against the yolk (cf. Fig. 103, A–F). As a result, the bent ventricular region begins to swing somewhat ventrad and lies less closely against the body of the embryo (Fig. 103, E, F).

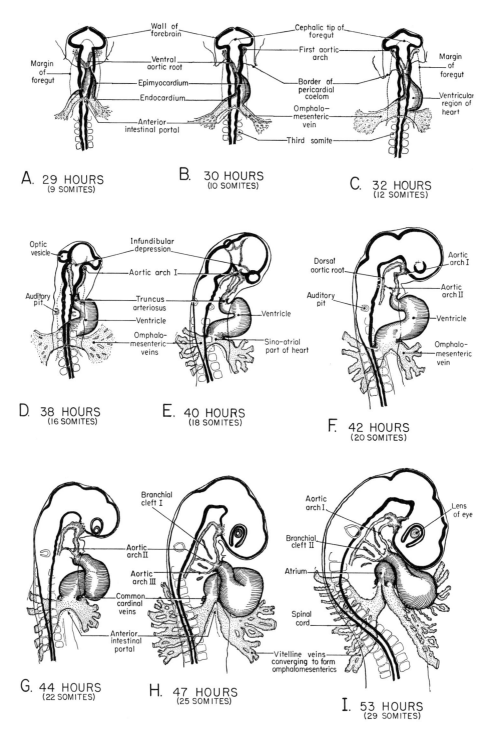

A. 29 HOURS (9 SOMITES)

B. 30 HOURS (10 SOMITES)

C. 32 HOURS (12 SOMITES)

D. 38 HOURS (16 SOMITES)

E. 40 HOURS (18 SOMITES)

F. 42 HOURS (20 SOMITES)

G. 44 HOURS (22 SOMITES)

H. 47 HOURS (25 SOMITES)

I. 53 HOURS (29 SOMITES)

Fig. 103. Projection drawings of the heart and great vessels of chick embryos of various ages. Some of the major topographical features of the embryos have been outlined to empha-

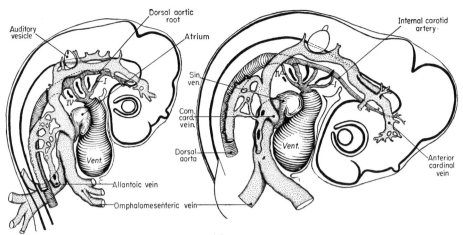

J. 65 HOURS (33 SOMITES)

K. 76 HOURS (38 SOMITES)

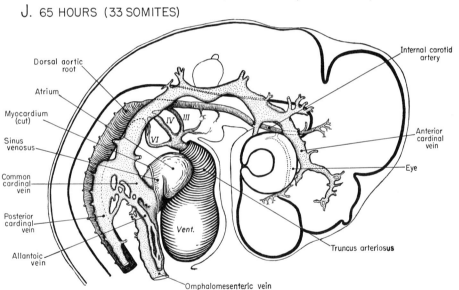

L. 100 HOURS (45 SOMITES)

size the relations of the heart within the body. The lettering of the parts of this figure corresponds with the more detailed drawings of the isolated hearts in Figs. 104 and 105.

Abbreviations: Roman numerals I–VI, aortic arches of the designated numbers; *Sin. ven.*, sinus venosus; *Vent.*, ventricle. [*Redrawn, slightly modified, from Patten, Am. J. Anat.,* **30** (1922).]

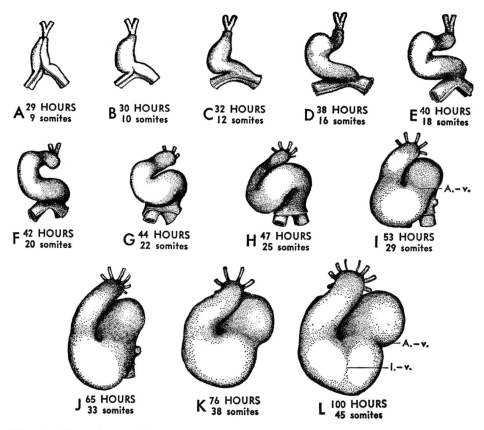

Fig. 104.   Ventral views of the heart at various stages to show its changes of shape and its regional differentiation. All the drawings were made from dissections with the aid of camera lucida outlines. In the stages represented in figures *E* to *H*, torsion of the embryo's body is going on at the level of the heart. Since torsion involves the more cephalic regions first and progresses caudad, the transverse axis of the body of the embryo is at different inclinations to the yolk at the cephalic end and at the caudal end of the heart. In drawing these figures their orientation was taken from the body at the level of the conus region of the heart; the sinus region therefore appears inclined.

   Abbreviations: A.-v., constriction between atrium and ventricle; *I.-v.,* interventricular groove. [*From Patten, Am. J. Anat.,* **30** (1922).]

At about this stage of development the closed part of the U-shaped bend becomes twisted on itself to form a loop (Figs. 103, 104, and 105 *F–I*). In the formation of the loop the atrial region is forced slightly to the left (i.e., toward the yolk) and the truncus arteriosus is thrown across the atrial region by being bent to the right (i.e., away from the yolk) and then dorso-caudad. The ventricular region constituting the closed end of the loop swings

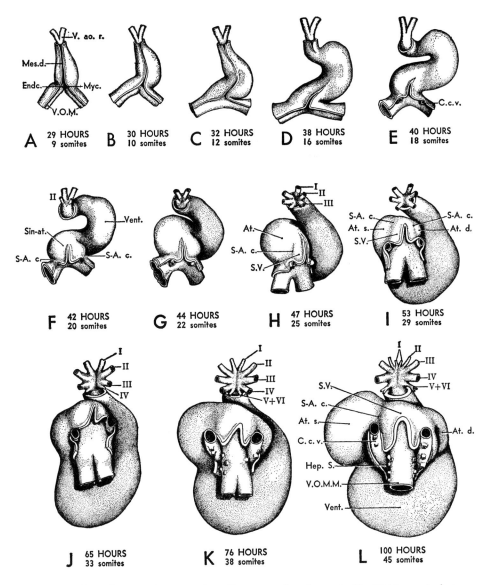

Fig. 105.   Dorsal views of same series of hearts as that shown in Fig. 104 in ventral view.

Abbreviations: I-VI, aortic arches I to VI; *At. d.,* atrium, right; *At. s.,* atrium, left; *C. c. v.,* common cardinal vein; *Endc.,* endocardium; Hep. S., stubs of some of the larger hepatic sinusoids; *Mes. d.,* dorsal mesocardium; *Myc.,* cut edge of epi-myocardium; *S-A. c.,* sino-atrial constriction; *Sin-at.,* sino-atrial region (before its definite division); *S. V.,* sinus venosus; *V. ao. r.,* ventral aortic roots; *Vent.,* ventricle; *V. O. M.,* omphalomesenteric veins; *V. O. M. M.,* fused omphalomesenteric veins. [*From Patten, Am. J. Anat.,* **30** (1922).]

dorsalward and toward the tail, possibly being crowded in this direction by the increasing flexion of the cephalic part of the embryo (Fig. 103, *G–J*). Thus the original cephalo-caudal relations of the atrial and ventricular regions are reversed, and the atrial region which was at first caudal to the ventricle now lies cephalic to it as it does in the adult heart.

The atrial region and the ventricular region, which formerly were continuous without any line of demarcation, are by this time beginning to be marked off from each other by a constriction [Fig. 104, *I* (a.v.)]. As both the atrium and the ventricle become enlarged, this constriction is accentuated [Fig. 104, *L* (a.v.)]. The constricted region is now termed the atrio-ventricular canal.

During the fourth day the truncus arteriosus becomes closely applied to the ventral surface of the atrium. As the atrium grows, it tends to expand on either side of the depression made in it by the pressure of the truncus. These lateral expansions are the first indication of the division of the atrium into right and left chambers which are later completely separated from each other. At the same time a slight longitudinal groove appears in the surface of the ventricle [Fig. 104, *L* (i.v.)]. This *interventricular groove* indicates the beginning of the separation of the ventricle into right and left chambers.

During the changes in the external shape of the heart which have been described, the whole heart has come to occupy a more caudal position with reference to other structures in the embryo. When the heart is first formed, it lies at the level of the rhombencephalon (Fig. 53). As development progresses, it moves farther and farther caudad until at the end of the fourth day the tip of the ventricle lies at the level of the anterior appendage-buds (Fig. 102).

During the third and fourth days of incubation, interesting changes become increasingly apparent in the ventricular portions of the cardiac wall. When the primordial tubular heart was first established, its endothelially lined lumen was smooth and fairly regular in diameter. Outside the endothelium, between it and the developing muscle, was a relatively thick layer of cardiac jelly (Fig. 59, *E*). Beginning in embryos of 50 to 55 hours (Fig. 106, *A*) and rapidly becoming more marked (Fig. 106, *B*), the myocardium shows irregular projections extending into the cardiac jelly. These projections are the start of the *trabeculae carneae* which are conspicuous features of the interior of the adult ventricles. As the trabeculae grow, the endothelial lining tends to extend between them and follow closely the configuration of the muscular strands separated from them only by a thin layer of cardiac jelly (Fig. 106, *C*).

Once established, the trabeculation shows an increasingly richly

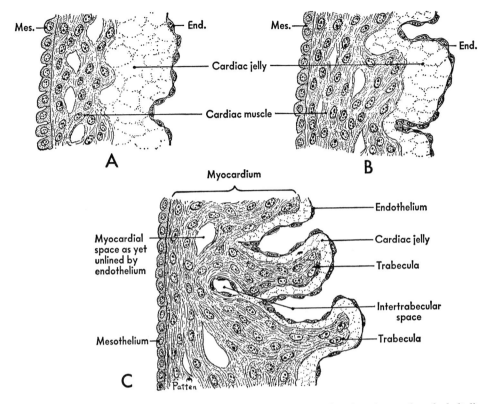

Fig. 106. High-power drawings to show the development of trabeculae and endothelially lined intertrabecular spaces in the ventricular walls of chick embryos. (*A*) At 50 hours; (*B*) at 65 hours; (*C*) fourth day.

branching pattern so that the ventricular wall becomes honeycombed by tortuous intertrabecular spaces, all of which are endothelially lined and all of which directly or indirectly communicate with the main lumen of the ventricle (Fig. 107). This structural pattern is of great functional significance, for it brings the blood in close relationship to the growing cardiac muscle during the period before the coronary circulation to the myocardium has been formed. In human embryos the heart is beating and effectively propelling the blood through the embryonic vessels for about a month before its coronary vessels develop. During that crucial period the myocardium is entirely dependent on the blood that reaches it by way of the intertrabecular spaces.

Even in embryos of the fourth day the primordial epicardium consists of but a single layer of cuboidal cells. These are the forerunners of the meso-

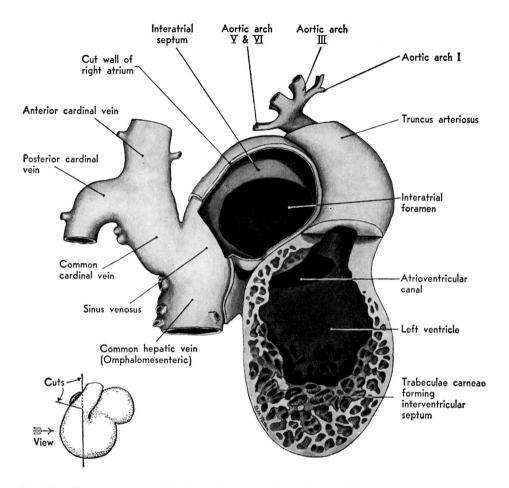

Interatrial septum

Aortic arch $\underline{V}$ & $\underline{VI}$

Aortic arch $\underline{III}$

Cut wall of right atrium

Aortic arch I

Anterior cardinal vein

Truncus arteriosus

Posterior cardinal vein

Interatrial foramen

Common cardinal vein

Atrioventricular canal

Sinus venosus

Left ventricle

Common hepatic vein (Omphalomesenteric)

Cuts

Trabeculae carneae forming interventricular septum

View

Fig. 107.   Reconstruction of the heart of a 4-day chick. The model is drawn as viewed from the left when cut open along the lines indicated in the orienting sketch. (*Patten and Zimmer.*)

thelium of the epicardium (Fig. 106, C). Later in development a supporting layer of epicardial connective tissue will be formed between the mesothelium and the underlying myocardium.

Lying between the endocardium and the myocardium in the region of the atrio-ventricular canal and of the opening of the ventricle into the truncus arteriosus, there are loosely aggregated masses of cells which resemble mesenchymal cells in their loose, sprawling appearance. There is evidence indicating that at least some of these cells arise originally from the endothelium instead of all of them coming from the myocardium, as has been commonly believed. Whatever their source they move into the space between the two primordial layers of the heart using the cardiac jelly as a sub-stratum on which to migrate. When there aggregated, they constitute what is called *endocardial cushion tissue*. This plastic connective tissue later takes part in the partitioning of the atrio-ventricular canal and in the formation of the connective tissue framework of the cardiac valves.

The definite establishment of the four-chambered condition character-istic of the adult heart belongs to later stages of development than those under consideration. By the end of the fourth day, however, the partitioning process is already begun. In reconstructions showing the interior of the heart the *interatrial septum* appears as a sickle-shaped partition cutting into the atrial lumen along the line where it is already narrowed by the pressure of the truncus arteriosus (Fig. 107). On the interior of the ventricular wall, opposite the inter-ventricular groove (Fig. 104, L), the trabeculae of growing muscle are especially abundant and project well into the ventricular lumen (Fig. 107). Consolidation of these trabeculae establishes a definite septum which grows from the apex of the ventricle toward the atrio-ventricular canal. Convergent growth of the interatrial and inter-ventricular septa, and their ultimate fusion with a partition concurrently established in the atrio-ventricular canal, finally accomplish the division of the heart into right and left sides. At the same time truncus arteriosus is divided into pulmonary and aortic channels communicating, respectively, with the right and left ven-tricle. With the completion of these processes the heart is prepared to pump separate pulmonary and systemic blood streams.

There are exceedingly interesting physiologic differences in the myo-cardium of these basic cardiac regions. It will be recalled that, in dealing with the establishing of the cardiac tube, emphasis was laid on the sequential manner of its formation. Only as the foregut acquires a floor can the cardiac primordia meet and fuse in the midline. That means that the truncoventricu-lar part of the heart is formed first, then fusion reaches the atrial region, and last of all the sinus venosus is established caudal to the atrium. Pulsation of the myocardium appears in these regions in the same sequence in which they are laid down. The first contractions appear in the ventricle before the atrium is fully established. The contraction rate of the ventricle is at first very slow. When the atrium begins to pulsate, the heart rate increases. We have already seen how transection experiments demonstrate that this in-crease in rate is due to the fact that the atrial myocardium has a faster in-herent rate of pulsation (Fig. 66) and takes control of cardiac rhythm as a pacemaker.

If now we carry out similar transection experiments after the sinus venosus has been formed caudal to the atrium, we find its rate of contraction

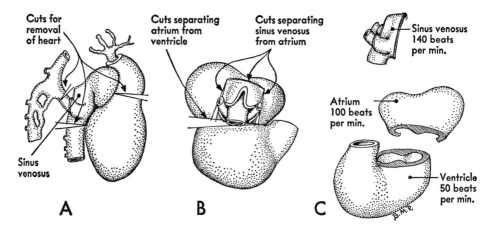

Fig. 108. Diagram showing the location of cuts made in the living heart of a 4-day chick embryo to demonstrate the relative pulsation rates of the myocardium from different regions. (*A*) Cuts for the removal of the heart. (*B*) Cuts to separate sinus, atrium, and ventricle. (*C*) The parts of the heart as isolated. The rates indicated are representative of the findings in such experiments rather than specific for any particular case. [*From Patten, Univ. of Mich. Medical Bull.*, **22** (1956).]

is higher than that of the atrium (Fig. 108). There is thus a cephalo-caudal gradient in the intrinsic rate of myocardial contraction. The importance of this situation is far-reaching. It means that the pacemaking center of the heart is, in all stages of development, at its intake end. Therefore, when a wave of contraction starts at this area and sweeps throughout the length of the cardiac tube, it picks up the incoming blood and forces it out at the other end of the heart into the vessels by which it will be distributed to the body. One would have great difficulty postulating a more efficient type of pumping action in a simple tubular heart without valves.

VI.   THE URINARY SYSTEM

General Relationship of Pronephros, Mesonephros, and Metanephros   In the development of the urinary system of birds and mammals there are formed in succession three distinct excretory organs, pronephros, mesonephros, and metanephros. The pronephros is the most anterior of the three and the first to be formed. It is wholly vestigial, appearing only as a slurred-over recapitulation of structural conditions which exist in the adults of the most primitive of the vertebrate stock. The mesonephros of a bird or mammalian embryo is homologous with the adult excretory organs of fishes and amphibia. It makes its appearance somewhat later than the pronephros, and is formed caudal to it. The mesonephros is the principal organ of excretion during early embryonic life, but it also disappears in the adult, except for parts of its duct system which become associated with the reproductive organs. The metanephros is the most caudally located of the excretory organs and the last to appear. It becomes functional toward the end of embryonic life when the mesonephros is disappearing, and persists permanently as the functional kidney of the adult.

Figure 109 shows schematically some of the main steps in the embryological history of the nephric organs, which it will be helpful to have in mind before taking up in detail any of the phases of their formation in the chick. The pronephros, mesonephros, and metanephros are all derived from the intermediate mesoderm, and are all composed of units which are tubular in nature. In the different nephroi these tubules vary in structural detail, but their functional significance is in all cases much the same. They are concerned in collecting waste materials from the capillary plexuses which are developed in connection with them. In the accompanying diagrams conventionalized tubules have been drawn to represent the three nephric organs. No pretense is made of representing either the exact shape or the actual number of the tubules.

In the first stage represented (Fig. 109, *A*) only the pronephros has been established. It consists of a group of tubules emptying into a common duct, called the *pronephric* or *primary nephric duct*.[1] The primary nephric ducts of either side are formed first at the level of the pronephric tubules and are then extended caudad, eventually reaching and opening into the cloaca. (See arrows in Fig. 109, *A*.)

As the primary nephric ducts are extended caudal to the level at which pronephric tubules are formed, they come in close proximity to the developing mesonephric tubules. In their growth the mesonephric tubules extend toward the primary nephric ducts and soon open into them (Fig. 109, *B*). Meanwhile the pronephric tubules begin to degenerate. Thus the ducts which originally arose in connection with the pronephros are appropriated by the developing mesonephros. After the degeneration of the pronephric tubules these same ducts are called the mesonephric ducts because of their new associations (Fig. 109, *C*).

---

[1] Students are frequently bothered by calling the same duct pronephric at one stage and mesonephric at another. To meet this difficulty we have here employed the term *primary nephric duct*, which seems to be less confusing. If this duct is not tagged with the specific name pronephric, it seems easier to explain its serving as the mesonephric duct after the pronephros has undergone regression.

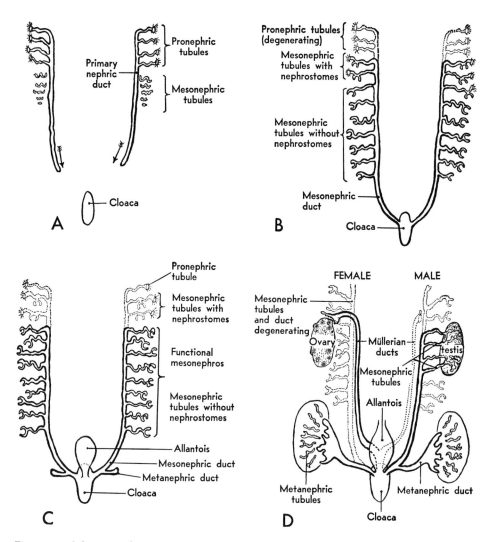

Fig. 109. Schematic diagrams to show the relations of pronephros, mesonephros, and metanephros at various stages of development. For the sake of simplicity the tubules have been drawn as if they had been pulled out to the side of the ducts. Their actual positional relations within the body are shown in Fig. 110.

The plan, as shown in *C*, represents approximately the conditions attained by the chick at the end of the fourth day of incubation or by the human embryo toward the end of the fourth week. In *D* are depicted the conditions after sexual differentiation has taken place; female left side, male right side of the diagram. The Müllerian ducts (shown in *D*) arise during the chick's fifth day of incubation in close association with the mesonephric ducts. They are the primordial tubes from which the oviducts of the female are formed. Note that while both mesonephric and Müllerian ducts appear in all young embryos, the Müllerian ducts become vestigial in the male and the mesonephric ducts become vestigial in the female.

At a considerably later stage outgrowths develop from the mesonephric ducts near their cloacal ends (Fig. 109, C). These outgrowths form the ducts of the metanephroi. They grow cephalo-laterad and eventually connect with the third group of tubules developed from the intermediate mesoderm, the metanephric tubules (Fig. 109, D). With the establishment of the metanephroi or permanent kidneys the mesonephroi begin to degenerate. The only parts of the mesonephric system to persist, except in vestigial form, are some of the ducts and tubules which, in the male, are appropriated by the testis as a duct system.

The Pronephric Tubules of the Chick   The pronephros in the chick is represented by tubules which first appear at about 36 hours of incubation. The pronephric tubules arise from the intermediate mesoderm, or nephrotome, lateral to the somites. They are paired, segmentally arranged structures, a tubule appearing on either side opposite each somite from the 5th to the 16th. Transverse sections passing through the 10th to 14th somites of an embryo of about 38 hours show the pronephric tubules favorably. Each tubule arises as a solid bud of cells organized from the intermediate mesoderm near its juction with the lateral mesoderm (Fig. 110, A). At first the free ends of the buds grow dorsad, passing close to the posterior cardinal veins. Later the end of each tubule is bent caudad coming in contact with the tubule lying posterior to it. In this manner the distal ends of the tubules give rise to a continuous cord of cells, the primordium of the primary nephric duct. The pair of cell cords thus formed continue to extend caudad beyond the pronephric tubules and soon become hollowed out to form open ducts. When they eventually reach the level of the cloaca they turn ventrad and open into it.

The significance of the rudimentary structures in the chick, which represent pronephric tubules, can most readily be understood by comparing them with fully developed and functional pronephric tubules. Figure 110, B, shows the scheme of organization of a functional pronephric tubule. The ciliated *nephrostome* draws in fluid from the coelom. As the coelomic fluid passes in close proximity to the capillaries of a vascular ridge known as the *glomus*, waste materials from the blood are transferred to it. The nephric duct serves to collect and discharge the fluid passing through the tubules. In the pronephric tubules of the chick there are vestiges of a nephrostome opening into the coelom (Fig. 110, A) but the tubules never become completely patent and never acquire the vascular relations characteristic of the functional pronephros in primitive vertebrates. Shortly after their initial appearance the pronephric tubules begin to undergo regressive changes and by the end of the fourth day of incubation a few isolated epithelial vesicles are all that remain to chronicle the transitory appearance of the pronephros.

The Mesonephric Tubules   The mesonephric tubules develop from the intermediate mesoderm caudal to the pronephros. The early steps in their formation are well shown in transverse sections of chicks of 29 to 30 somites (about 55 hours). In the posterior region conditions are less advanced than they are more anteriorly. Consequently by studying the posterior sections of a transverse series first and then progressing cephalad, a graded series of developmental stages may be obtained.

The mesonephric tubules appear first as cell clusters formed in the intermediate mesoderm ventro-mesial to the cord of cells which is the primordium of the primary nephric duct. The cells of the developing tubules acquire a more or less radial arrangment, and at the same time become more distinctly isolated from the surrounding mesodermal cells. By 55 hours of incubation the primordial cell cord representing the primary nephric duct has become hollowed out to establish a definite lumen. The most anterior of the mesonephric tubules also have acquired a lumen. Meanwhile the growth of the tubules has brought them in close association with the duct and in some of the more differentiated tubules indications can be made out of their opening into the duct. The more posterior mesonephric tubules do not become associated with the duct until somewhat later, but remain as a series of isolated vesicles.

The mesonephric tubule differs from the pronephric chiefly in its relation to the blood vessels associated with it. It develops a cup-like outgrowth into which a knot of capillaries is pushed. The cup-shaped outgrowth from the tubule is called the *glomerular capsule*, or *capsule of Bowman*, and the tuft of capillaries is termed a *glomerulus*. The diminutive form *glomerulus* is used to distinguish such small, localized capilary tufts from the continuous vascular ridge called a *glomus*. From the capillaries of the glomerulus there is free filtration of the fluid portion of the blood into the glomerular capsule. Then as this fluid passes along the lumen of the tubule there is selective resorption into the adjacent capillaries of such materials as salts and sugars, together with a large amount of water. It is this regulated resorption of substances in the nephric tubules that is responsible for the maintenance of the delicate salt-water balance which exists in the blood and tissue fluids. In this process waste materials are left within the tubule and carried by the balance of the water into the excretory duct and thence into the allantois for storage. The resorption of the great bulk of the water that originally entered the

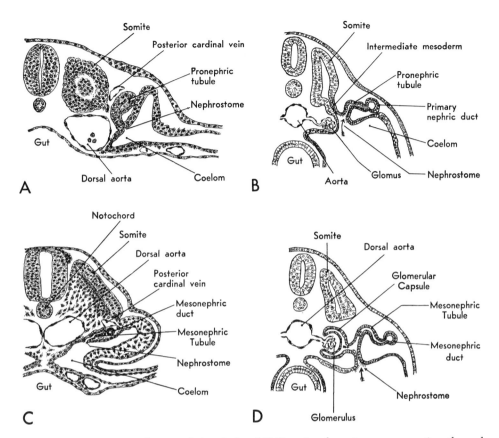

Fig. 110. Drawings to show nephric tubules. (*A*) Drawing from transverse section through twelfth somite of 16-somite chick to show pronephric tubule. (*After Lillie.*) (*B*) Schematic diagram of functional pronephric tubule. (*After Wiedersheim.*) (*C*) Drawing from transverse section through seventeenth somite of 30-somite chick to show primitive mesonephric tubule. (*D*) Schematic diagram of functional mesonephric tubule of primitive type. (*After Wiedersheim.*)

tubule from the glomerulus conserves water for the body and at the same time concentrates the waste materials.

In a mesonephric tubule of the more primitive type such as that diagrammed in Fig. 110, *D* there is a ciliated nephrostome which draws in fluid from the coelom. This fluid mingles with the fluid from the glomerulus and is subjected to the same selective resorption in the nephric tubule. In the embryos of higher vertebrates only a few of the more cephalic of the mesonephric tubules ever show nephrostomes, and these are rudimentary (Fig. 110, *C*). The great bulk of the functioning mesonephros is made up of tubules which form no nephrostome (Fig. 111) but depend entirely on their glomerular apparatus for their fluid intake.

In chicks of four days incubation the mesonephric tubules have not attained their full development, but it is possible to make out most of their fundamental parts (Fig. 111). The tubules lying in the more cephalic part of the mesonephros have been longest established and are somewhat farther advanced in development than those lying more caudally in the mesonephros. Nearly all of the tubules, however, have become elongated and somewhat coiled. At one end they open into the mesonephric duct which acts as a main excretory channel. At their other end a cluster of closely packed cells, which is beginning to take on a cup-shaped arrangement, indicates the place at which the capsule will appear. The glomerulus is suggested by small vessels, or by endothelial sprouts, becoming organized in the concavity of the developing capsule. Once established, the glomeruli develop very rapidly. Circulation usually commences in them by the fifth day. From this time until about the eleventh day of incubation the functional activity of the

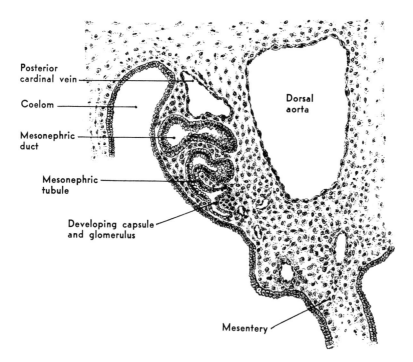

Fig. 111.   Drawing from transverse section of 4-day chick to show mesonephric tubule and duct. For the location of the area drawn, consult Fig. 97, F.

mesonephros is at its height. After the eleventh day the developing meta-nephros begins to become active and the mesonephros degenerates. In discussing the distal portion of the allantois as a place of storage of waste materials, mention was made of the change in excretory product which occurs as development progresses. It will be recalled that urea is the predominant nitrogenous waste product of early stages and that uric acid is the chief form in later stages. The importance of this change lies in the low solubility of uric acid and the consequent lessening of the problem of its storage without toxic effects. This change in the character of the excretory products occurs as the metanephros takes over the excretory function from the regressing mesonephros.

The Metanephros   The differentiation of the metanephros and the development of the genital organs occur in stages which are too advanced to come within the scope of this book. Young mammalian embryos such as those of the pig or rabbit make excellent laboratory material for the study of these more advanced phases of the development of the urogenital system and are widely used for this purpose. Those interested in studying this system as it develops in older bird embryos may consult a reference book such as Lillie's "The Development of the Chick."

VII.   THE COELOM AND MESENTERIES

In adult birds and mammals the body-cavity consists of three regions, pericardial, pleural, and peritoneal. The *pleural cavities* are paired, each of the pleural chambers being a laterally situated sac containing one of the lungs. The *pericardial chamber* containing the heart and the *peritoneal chamber* containing the viscera other than the lungs and heart are unpaired. These regions of the adult body-cavity are formed by the partitioning off of the primary body-cavity or coelom of the embryo.

In the chick the coelom arises by a splitting of the lateral mesoderm on either side of the body (Fig. 112, *A*, *B*). It is, therefore, primarily a paired cavity. Unlike the coelom of some of the more primitive vertebrates, the coelom of the chick never shows any indications of segmental pouches corresponding in arrangement with the somites. The right and left coelomic chambers extend cephalo-caudally without interruption through the entire lateral plates of mesoderm. This difference in the formation of the coelom does not imply any lack of homology between the coelom of the chick and that of

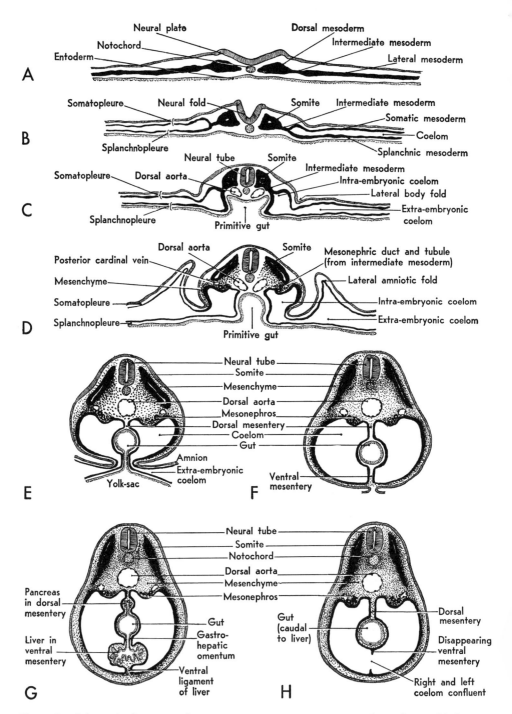

Fig. 112. Schematic diagrams of cross sections at various stages to show the establishment of the coelom and mesenteries. For explanation see text.

more primitive forms. The process of coelom formation in the chick may be considered as being accelerated with a resultant slurring-over of the early phases. Or, to state the matter in another way, the coelom of birds and mammals first appears in a condition which is comparable with the coelom of more primitive forms at that period of differentiation when the segmentally arranged coelomic pouches have broken through into each other and their cavities have become confluent.

The coelomic chambers are not limited to the region in which the body of the embryo is developing. They extend on either side into the mesoderm, which, in common with the other germ layers, spreads out over the yolk surface. Large parts of the primitive coelomic chambers thus come to be extraembryonic in their associations. (See Chap. 10 and Figs. 68 and 71.) The portion of the coelom which gives rise to the embryonic body cavities is first marked off by the series of folds which separate the body of the embryo from the yolk (Fig. 112, C, D). As the closure of the ventral body-wall progresses (Fig. 112, E, F), the embryonic coelom becomes completely separate from the extra-embryonic. The delayed closure of the ventral body-wall in the yolk-stalk region results in the embryonic and extra-embryonic coelom retaining their open communication at this point for a considerable time after they have been completely separated elsewhere.

The same folding process which establishes the ventral body-wall completes the gut ventrally (Fig. 101, C–F). Meanwhile the right and left coelomic chambers are expanded mesiad. As a result the newly closed gut comes to lie suspended between the two layers of splanchnic mesoderm which constitute the mesial walls of the right and left coelomic chambers. The double layer of splanchnic mesoderm which thus becomes apposed to the gut and supports it in the body-cavity is known as the *primary mesentery*. The part of the mesentery dorsal to the gut, suspending it from the dorsal body-wall, is the *dorsal mesentery*, and the part ventral to the gut, attaching it to the ventral body-wall, is the *ventral mesentery*.

When the dorsal and ventral mesenteries are first established, they constitute a complete membranous partition dividing the body-cavity into right and left halves. The primary dorsal mesentery persists in large part, but the ventral mesentery early disappears (Fig. 112, H), bringing the right and left coelomic chambers into confluence ventral to the gut and establishing the unpaired condition of the body-cavity characteristic of the adult.

In considering the early development of the heart (Chap. 8), the formation of the dorsal and ventral mesocardia was discussed. In their relation to the other mesenteries of the body, the mesocardia may be regarded as special regions of the ventral mesentery. In the most cephalic part of the body-

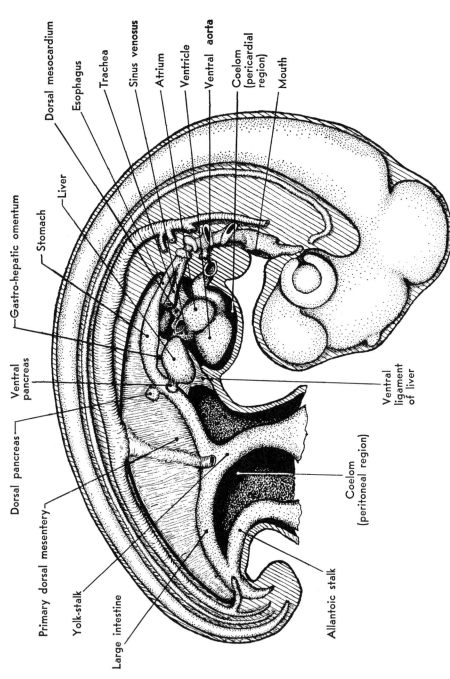

Dorsal mesocardium

Esophagus

Trachea

Sinus venosus

Atrium

Ventricle

Ventral aorta

Coelom
(pericardial
region)

Mouth

Liver

Gastro-hepatic omentum

Stomach

Ventral pancreas

Dorsal pancreas

Primary dorsal mesentery

Yolk-stalk

Large intestine

Allantoic stalk

Coelom
(peritoneal region)

Ventral
ligament
of liver

Fig. 113. Schematic lateral view of dissection of 4-day chick to show the various parts of the body-cavity and the more important mesenteries.

cavity, the gut lies embedded in the dorsal body-wall instead of being suspended by a dorsal mesentery as it is farther caudally (cf. Figs. 59, E, and
112, F). A ventral mesentery is, however, developed in the same manner
anteriorly as it is posteriorly, and when the heart is formed, it is suspended
in the most anterior part of this ventral mesentery. The dorsal and ventral
mesocardia may, therefore, be thought of as the parts of the primary ventral
mesentery lying dorsal to the heart and ventral to the heart, respectively (Fig.
59, D).

When the ventral mesocardium, and a little later the dorsal mesocardium, break through, the primary right and left coelomic chambers become
confluent to form the pericardial region of the body-cavity (Figs. 64 and 113).
Later in development the ventral mesentery farther caudally disappears, so
that caudally as well as cephalically an unpaired condition of the coelom is
brought about (Fig. 112, H).

In the liver region a portion of the ventral mesentery remains intact.
The liver arises as an outgrowth from the gut and in its development extends
into this retained part of the ventral mesentery (Fig. 112, G). That portion of
the ventral mesentery which is dorsal to the liver persists as the *gastro-
hepatic omentum*, and that portion of the mesentery which is ventral to the
liver persists as the *falciform ligament or ventral ligament* of the liver (Fig.
113).

The entire dorsal mesentery persists and forms the supporting membranes of the digestive tube. In the adult its different regions are named according to the parts of the digestive tube with which they are associated, as
for example, mesogaster, that part of the dorsal mesentery which suspends
the stomach; mesocolon, that part of the dorsal mesentery supporting the
colon, etc.

The separation of the body-cavity into pericardial, pleural, and
peritoneal chambers is accomplished by the formation of septa growing in
from the body-wall. Consideration of the details of their formation would
lead us into stages of development beyond the scope of this book. Those interested in following this or other phases of the later embryology of the chick
will find in the bibliography references to more exhaustive books, and to
some of the more recent original papers on its development.

BIBLIOGRAPHY

Although this does not purport to be a complete bibliography on the embryology of the chick, I have attempted to make it reasonably comprehensive. Many references which involve phases of the subject not dealt with in this elementary text have been included. At the same time an effort has been made to keep the list within reasonable bounds by limiting it largely to recent articles in readily accessible publications and omitting references dealing primarily with mammalian development. With the exception of a few classic contributions, papers of merely historical interest have not been included. Those interested in this aspect of the subject can find, under each of the main headings, articles having extensive bibliographies covering the old as well as the current literature in their special fields. It is hoped that the instructor will find the references a convenience in selecting collateral reading, and especially that the student who becomes spontaneously interested in knowing more about some special phase of the subject can find references which will start him toward acquiring the information he desires.

For convenience in locating material the references have been grouped under the following headings:

I   Texts and General Treatises
II Gametogenesis and Fertilization
III Cleavage, Gastrulation, and Germ Layer Formation
IV Early Differentiation of the Embryo

I. Texts and General Treatises

Balinsky, B. I., 1970. An Introduction to Embryology. W. B. Saunders Company, Philadelphia. 3d ed., xviii + 725 pp.

Barth, L. G., 1953. Embryology. Holt, Rinehart and Winston, Inc., New York. 2d ed, xi + 516 pp.

Duval, M., 1889. Atlas d'embryologie. Masson, Paris, 116 pp., 40 plates.

Hertwig, O., 1901–1907. Handbuch der vergleichenden und experimentellen Entwickelungslehre der Wirbeltiere. (Edited by Hertwig, written by numerous collaborators.) Fischer, Jena.

Huettner, Alfred F., 1949. Fundamentals of Comparative Embryology of the Vertebrates. The Macmillan Company, New York. 2d ed., xviii + 309 pp.

Kaupp, B. F., 1918. The Anatomy of the Domestic Fowl. W. B. Saunders Company, Philadelphia. 373 pp.

Keibel, F., and Abraham, K., 1900. Normentafeln zur Entwickelungsgeschichte des Huhnes (Gallus domesticus). Fischer, Jena, 132 pp., 3 plates.

Lillie, F. R., 1952. The Development of the Chick. Henry Holt and Company, New York. 3d ed., xv and 624 pp., edited by Howard L. Hamilton.

Long, J. A., and Burlingame, P. L., 1937. A Stereoscopic Atlas of the Chick. California Laboratory Supply Co., Los Angeles.

McEwen, R. S., 1949. Vertebrate Embryology. Henry Holt and Company, New York, 3d ed., xv and 699 pp.

Minot, C. S., 1893. A bibliography of vertebrate embryology. Memoirs, Boston Soc. Nat. History, 4:487–614.

Needham, J., 1931. Chemical Embryology. The Macmillan Company, New York. xxii + 2021 pp.

Patten, B. M., 1964. Foundations of Embryology. McGraw-Hill Book Company, New York. 2d ed., xix and 622 pp.

Romanoff, A. L., 1960. The Avian Embryo. Structural and Functional Development. The Macmillan Company, New York. 1305 pp.

———and Romanoff, A. J. S., 1967. Biochemistry of the Avian Embryo. A quantitative analysis of prenatal development. John Wiley & Sons, Inc., New York. 398 pp.

Rugh, R., 1964. Vertebrate Embryology. Harcourt, Brace and World, Inc., New York. viii and 600 pp.

Spemann, H., 1938. Embryonic Development and Induction. Yale University Press, New Haven, Conn. xii + 401 pp.

Waddington, C. H., 1956. Principles of Embryology. The Macmillan Company, New York. 3d printing, 1960, x + 510 pp.

Witschi, E., 1956. Development of Vertebrates. W. B. Saunders Company, Philadelphia. xiv + 588 pp.

II. Gametogenesis and Fertilization

Asmundson, V. S., and Burmester, B. R., 1936. The secretory activity of the parts of the hen's oviduct. J. Exptl. Zool., 72:225–246.

Bartelmez, G. W., 1912. The bilaterality of the pigeon's egg. A study in egg organization from the first growth period of the oocyte to the beginning of cleavage. J. Morphol., 23:269–328.

Burmester, B. R., 1940. A study of the physical and chemical changes of the egg during its passage through the isthmus and uterus of the hen's oviduct. J. Exptl. Zool., 84:445–500.

Carothers, E. E., 1926. The maturation divisions in relation to the segregation of homologous chromosomes. Quart. Rev. Biol., 1:419–435.

Cole, R. K., 1938. Histology of the oviduct of the fowl in relation to variations in the condition of the firm egg albumen. Anat. Record, 71:349–361.

Conrad, R. M., and Scott, H. M., 1938. The formation of the egg in the domestic fowl. Physiol. Rev., 18:481–494.

Curtis, M., 1915. Relation of simultaneous ovulation to the production of double-yolked eggs. J. Agr. Res., 3:375–385.

Davis, D. E., 1942a. The bursting of avian follicles at the beginning of atresia. Anat. Record, 82:153–165.

———, 1942b. The regression of the avian postovulatory follicle. Anat Record, 82:297–307.

Fröböse, H., 1928. Die mikroskopische Anatomie des Legedarmes und Bemerkungen über die Bildung der Kalkschalle beim Huhn. Zeitschr. f. mikr.-anat. Forsch., 14:447–482.

Glaser, O., 1913. On the origin of double-yolked eggs. Biol. Bull., 24:175–186.

Goldsmith, J. B., 1928. The history of the germ cells in the domestic fowl. J. Morphol. Physiol., 46:275–315.

Hance, R. T., 1926. Sex and the chromosomes in the domestic fowl (Gallus domesticus). J. Morphol. Physiol., 43:119–145.

Kemp, N. E., 1956. Electron microscopy of growing oocytes of Rana pipiens. J. Biophys. Biochem. Cytol., 2:281–292.

Knight, P. F., and Schechtman, A. M., 1954. The passage of heterologous serum proteins from the circulation into the ovum of the fowl. J. Exptl. Zool., 127:271–304.

Lillie, F. R., 1919. Problems of Fertilization. University of Chicago Press, xii + 278 pp.

Morgan, T. H., 1926. The Theory of the Gene. Yale University Press, New Haven, Conn. xiv + 343 pp.

Olsen, M. W., and Fraps, R. M., 1950. Maturation changes in the hen's ovum. J. Exptl. Zool., 114:475–490.

Patten, W., 1925. Life, evolution and heredity. Sci. Monthly, 21:122–134.

Pearl R., and Curtis, M. R., 1912. Studies on the physiology of reproduction in the domestic fowl. V. Data regarding the physiology of the oviduct. J. Exptl. Zool., 12:99–132.

_____ and Schoppe, W. F., 1921. Studies on the physiology of reproduction in the domestic fowl. XVIII. Further observations on the anatomical basis of fecundity. J. Exptl. Zool., 34:101–118.

Phillips, R. E., and Warren, D. C., 1937. Observations concerning the mechanics of ovulation in the fowl. J. Exptl. Zool., 76:117–136.

Purkinje, J. E., 1943. Contributions to the history of the bird's egg previous to incubation. Essays in Biology, University of California Press, pp. 55–92.

Riddle, O., 1911. On the formation, significance and chemistry of the white and yellow yolk of ova. J. Morphol., 22:455–492.

Romanoff, A. L., 1943. Growth of avian ovum. Anat. Record, 85:261–267.

_____ and Hutt, F. B., 1945. New data on the origin of double avian eggs. Anat. Record, 91:143–154.

Steggerda, M., 1928. Effect of ovarian injury on egg laying in fowls. J. Exptl. Zool., 51:403–416.

Swift, C. H., 1914. Origin and early history of the primordial germ-cells in the chick. Am. J. Anat., 15:483-516.

_____, 1915. Origin of the definitive sex-cells in the female chick and their relation to the primordial germ-cells. Am. J. Anat., 18:441–470.

_____, 1916. Origin of the sex-cords and definitive spermatogonia in the male chick. Am. J. Anat., 20:375–410.

Swingle, W. W., 1926. The determination of sex in animals. Physiol. Rev., 6:28–61.

Warren, D. C., and Scott, H. M., 1936. Influence of light on ovulation in the fowl. J. Exptl. Zool., 74:137–156.

Wilson, E. B., 1925. The Cell in Development and Heredity. The Macmillan Company, New York. 3d ed., iv + 1232 pp.

III.   Cleavage, Gastrulation and Germ Layer Formation

Butler, E., 1935. The developmental capacity of regions of the unincubated chick blastoderm as tested in chorio-allantoic grafts. J. Exptl. Zool., 70:357–395.

Duval, M., 1884. De la formation du blastoderme dans l'oeuf d'oiseau. Ann. Sci. Nat. Zool., 18:1–208.

Edwards, C. L., 1902. The physiological zero and the index of development for the egg of the domestic fowl. Am. J. Physiol., 6:351–397.

Fraser, R. C., 1954. Studies on the hypoblast of the young chick embryo. J. Exptl. Zool., 126:340–400.

Harman, M. T., 1922. Concerning the origin of the notochord in the chick. Anat. Record, 23:363–369.

Holtfreter, J., 1943–44. A study of the mechanics of gastrulation. I. J. Exptl. Zool., 94:261–318; II. J. Exptl. Zool., 95:171–212.

———, 1947. Changes of structure and the kinetics of differentiating embryonic cells. J. Morphol., 80:57–92.

Hunt, T. E., 1937. The origin of entodermal cells from the primitive streak of the chick embryo. Anat. Record, 68:449–459.

Jacobson, W., 1938a. The early development of the avian embryo. I. Endoderm formation. J. Morphol., 62:415–443.

———, 1938b. The early development of the avian embryo. II. Mesoderm formation and the distribution of presumptive embryonic material. J. Morphol., 62:445–501.

Just, E. E., 1912. The relation of the first cleavage plane to the entrance point of the sperm. Biol. Bull., 22:239–252.

Lewis, W. H., 1947. Mechanics of invagination. Anat. Record, 97:139–156.

Lutz, H., 1955. Contribution experimentale a l'étude de la formation de l'endoblaste chez les oiseaux. J. Embryol. Exptl. Morphol., 3(part I):59–79.

McWhorter, J. E., and Whipple, A. C., 1912. The development of the blastoderm of the chick in vitro. Anat. Record, 6:121–140.

Olsen, M. W., 1942. Maturation, fertilization, and early cleavage in the hen's egg. J. Morphol, 70:513–533.

Pasteels, J., 1937. Études sur la gastrulation des vertèbres méroblastiques. III. Oiseaux. IV. Conclusions générales. Arch. Biol., 48:381–488.

———, 1945. On the formation of the primary entoderm of the duck (Anas domestica) and on the significance of the bilaminar embryo in birds. Anat. Record, 93:5–21.

Patterson, J. T., 1909. Gastrulation in the pigeon's egg; a morphological and experimental study. J. Morphol., 20:65–123.

———, 1910. Studies on the early development of the hen's egg. I. History of the early cleavage and of the accessory cleavage. J. Morphol., 21:101–134.

Peter, K., 1938. Untersuchungen uber die Entwicklung des Dotterentoderms. I. Die

Entwicklung des Entoderms beim Hühnchen. Zeitschr. f. mikr.-anat. Forsch., **43**:362–415.

Rudnick, D., 1944. Early history and mechanics of the chick blastoderm. Quart. Rev. Biol., **19**:187–212.

Vogt, W., 1929. Gestaltungsanalyse am Amphibienkeim mit örtlicher Vitalfärbung. II. Gastrulation und Mesodermbildung bei Urodelen und Anuren. Arch. Entw.-mech. Organ., **120**:384–706.

Wetzel, R., 1925. Untersuchungen am Hühnerkeim. I. Über die Untersuchung des lebenden Keims mit neueren Methoden, besonders der Vogtschen vitalen Farbmarkierung. Arch. Entw.-mech. Organ., **106**:463–468.

———, 1929. Untersuchungen am Hühnchen. Die Entwicklung des Keims während der ersten beiden Bruttage. Arch. Entw.-mech. Organ., **119**:188–321.

Young, R. T., 1923. Origin of the notochord in chordates. Anat. Record, **25**:289–290.

IV.   Early Differentiation of the Embryo

Abercrombie, M., 1950. The effects of antero-posterior reversal of lengths of the primitive streak in the chick, Phil. Trans. Royal Soc. London, **234**:317–338.

Adelmann, H. B., 1922. The significance of the prechordal plate: An interpretative study. Am. J. Anat., **31**:55–101.

———, 1926. The development of the premandibular head cavities and the relations of the anterior end of the notochord in the chick and robin. J. Morphol. Physiol., **42**:371–439.

Allen, H. J., 1919. Glycogen in the chick embryo. Biol. Bull., **36**:63–70.

Andres, G., 1955. Growth reactions of mesonephros and liver to intravascular injections of embryonic liver and kidney suspensions in the chick embryo, J. Exptl. Zool., **130**:221–249.

Bartelmez, G. W., 1918. The relation of the embryo to the principal axis of symmetry in the bird's egg. Biol. Bull., **35**:319–361.

Boyden, E. A., 1918. Vestigial gill filaments in chick embryos with a note on similar structures in reptiles. Am. J. Anat., **23**:205–235.

Buno, W., 1951. Localization of sulfhydryl groups in the chick embryo. Anat. Record, **111**:123–128.

Cairns, J. M., and Saunders, J. W., Jr., 1954. The influence of embryonic mesoderm on the regional specification of epidermal derivatives in the chick. J. Exptl. Zool., **127**:221–248.

Clarke, L. F., 1936. Regional differences in eye-forming capacity of the early chick blastoderm as studied in chorio-allantoic grafts. Physiol. Zool., **9**:102–128.

Dalton, A. J., 1935. The potencies of portions of young chick blastoderms as tested in chorio-allantoic grafts. J. Exptl. Zool., **71**:17–51.

DeHaan, R. L., and Ebert, J. D., 1964. Morphogenesis, Ann. Rev. Phys., **26**:15–46.

Derrick, G. E., 1937. An analysis of the early development of the chick by means of the mitotic index. J. Morphol., **61**:257–284.

Fraser, R. C., 1960. Somite genesis in the chick. III. The role of induction. J. Exptl. Zool., **145**:151–167.

Gaertner, R. A., 1949. Development of the posterior trunk and tail of the chick embryo. J. Exptl. Zool., **111**:157–174.

Gluecksohn-Waelsch, S., 1954. Genetic control of embryonic growth and differentiation. J. Natl. Cancer Inst., **15**:629–634.

Grabowski, C. T., 1956. The effects of the excision of Hensen's node on the early development of the chick embryo. J. Exptl. Zool., **133**:301–344.

———, 1962. Neural induction and notochord formation by mesoderm from the node area of the early chick blastoderm. J. Exptl. Zool., **150**:233–246.

Gruenwald, P., 1941. Normal and abnormal detachment of body and gut from the blastoderm in the chick embryo, with remarks on the early development of the allantois, J. Morphol., **69**:83–125.

Hamburger, V., and Hamilton, H. L., 1951. A series of normal stages in the development of the chick embryo. J. Morphol., **88**:49–92.

Hansborough, L. A., and Nicholas, P. A., 1949. The distribution of phosphorus in the early chick embryo in the presence of vitamin D. J. Exptl. Zool., **112**:195–205.

Harrison, R. G., 1914. The reaction of embryonic cells to solid structures, J. Exptl. Zool., **17**:521–544.

Herrmann, H., Schneider, M. J. B., Neukom, B. J., and Moore, J. A., 1951. Quantitative data on the growth process of the somites of the chick embryo: Linear measurements, volume, protein nitrogen, nucleic acids. J. Exptl. Zool., **118**: 243–268.

Hinsch, G. W., and Hamilton, H. L., 1956. The developmental fate of the first somite of the chick. Anat. Record, **125**:225–246.

His, W., 1868. Untersuchungen über die erste Anlage des Wirbelthierleibes. Vogel, Leipzig. xvi + 237 pp.

Hoadley, L., 1924–1925. The independent differentiation of isolated chick primordia in chorio-allantoic grafts.

    I. The eye, nasal region, otic region, and mesencephalon. Biol. Bull. **46**: 231–315.

    II. The effect of the presence of the spinal cord, i.e., innervation, on the differentiation of the somitic region. J. Exptl. Zool., **42**:143–162.

    III. On the specificity of nerve processes arising from the mesencephalon in grafts. J. Exptl. Zool., **42**:163–182.

———, 1926. Developmental potencies of parts of the early blastoderm of the chick.

    I. The first appearance of the eye. J. Exptl. Zool., **43**:151–178.

    II. The epidermis and the feather primordia. J. Exptl. Zool., **43**:179–196.

III. The nephros, with especial reference to the pro- and mesonephric portions. J. Exptl. Zool., **43**:197–223.

————, 1927. Concerning the organization of potential areas in the chick blastoderm. J. Exptl. Zool., **48**:459–473.

Holmdahl, D. E., 1935. Primitivstreifen beziehungsweise Rumpfschwanzknospe im Verhaltnis zur Korperentwicklung. Zeitschr. f. mikr.-anat. Forsch., **38**:409–440.

————, 1939. Die formalen Verhältnisse während der Entwicklung der Rumpfschwanzknospe beim Huhn. Verhandel. Anat. Ges., Ergänzungsheft Anat. Anz., **88**:127–137.

Hubbard, M. E., 1908. Some experiments on the order of succession of the somites of the chick. Am. Naturalist, **42**:466–471.

Hughes, A. F. W., 1934. On the development of the blood vessels in the head of the chick. Phil. Trans. Roy. Soc. London, Ser. B, **224**:75–161.

Hunt, E. L., and Wolken, J. J., 1948. The distribution of phosphorus in early chick embryos. J. Exptl. Zool., **109**:109–118.

Hunt, T. E., 1931. An experimental study of the independent differentiation of the isolated Hensen's node and its relation to the formation of axial and non-axial parts in the chick embryo. J. Exptl. Zool., **59**:395–427.

————, 1932. Potencies of transverse levels of the chick blastoderm in the definitive-streak stage. Anat. Record, **55**:41–69.

————, 1937. The development of gut and its derivatives from the mesectoderm and mesentoderm of early chick blastoderms. Anat. Record, **68**:349–369.

Hurd, M. C., 1928. Observations on the storage of trypan blue in the embryo chick. Am. J. Anat., **42**:155–179.

Hyman, L. G., 1927. The metabolic gradients of vertebrate embryos. III, The chick. Biol. Bull., **52**:1–38.

Katz, B., 1961. How cells communicate, Sci. Am., **205**:209–220.

Kingsbury, B. F., 1926. On the so-called law of anteroposterior development. Anat. Record, **33**:73–87.

Lee, W. H., 1951. The glycogen content of various tissues of the chick embryo. Anat. Record, **110**:465–474.

McKeehan, M. S., 1958. Induction of portions of the chick lens without contact with the optic cup. Anat. Record, **132**:297–305.

Murray, H. A., 1925–26. Physiological ontogeny. A. Chicken embryo. (A series of articles on biochemical processes in the embryo.) J. Gen. Physiol., vols. 9 and 10.

Oppenheimer, J., 1959. Intercellular activities in vertebrate development. Science, **130**:686–692.

Patterson, J. T., 1907. The order of appearance of the anterior somites in the chick. Biol. Bull., **13**:121–133.

Peebles, F., 1904. The location of the chick embryo upon the blastoderm. J. Exptl. Zool., **1**:369–384.

Philips, F. S., 1942. Comparison of the respiratory rates of different regions of the chick blastoderm during early stages of development. J. Exptl. Zool., **90**:83–100.

Rao, B. R., 1968. The appearance and extension of neural differentiation tendencies in the neurectoderm of the early chick embryo. Roux's Arch. Entro.-mech. Organ., **160**:187–236.

Rosenquist, G. C., 1966. A radioautographic study of labeled grafts in the chick blastoderm. Development from primitive-streak stages to stage 12, Carnegie Contrib. Embryol., **38**:71–110.

Rudnick, D., 1932. Thyroid-forming potencies of the early chick blastoderm. J. Exptl. Zool., **62**:287–317.

_____, 1933. Developmental capacities of the chick lung in chorioallantoic grafts. J. Exptl. Zool., **66**:125–153.

_____, 1935. Regional restriction of potencies in the chick during embryogenesis. J. Exptl. Zool., **71**:83–99.

_____, 1945. Limb-forming potencies of the chick blastoderm: including notes on associated trunk structures. Trans. Conn. Acad. Arts Sci., **36**:353–377.

_____, 1967. Localization of glutamotransferase activity in the chick embryo during the first six days of incubation, Estratto dall'Archivio Zoologico Italiano, **51**:137–147.

Saunders, J. W., 1948. The proximo-distal sequence of origin of the parts of the chick wing and the role of the ectoderm. J. Exptl. Zool., **108**:363–403.

_____, Cairns, J. M., and Gasseling, M. T., 1957. The role of the apical ridge of ectoderm in the differentiation of the morphological structure and inductive specificity of limb parts in the chick. J. Morphol., **101**:57–87.

_____ and Gasseling, M. T., 1963. Trans-filter propagation of apical ectoderm maintenance factor in the chick embryo wing bud. Develop. Biol., **7**:64–78.

_____, Gasseling, M. T., and Saunders, L. C., 1962. Cellular death in morphogenesis of the avian wing. Develop. Biol., **5**:147–178.

Schmalhausen, J., 1926. Studien über Wachstum und Differenzierung. III. Die embryonale Wachstumskurve des Hühnchens. Roux's Arch. Entw.-mech. Organ., **108**:322–387.

Seno, T., 1961. An experimental study on the formation of the body wall in the chick. Acta Anat., **45**:60–82.

Spratt, N. T., Jr., 1942. Location of organ-specific regions and their relationship to the development of the primitive streak in the early chick blastoderm. J. Exptl. Zool., **89**:69–101.

_____, 1946. Formation of the primitive streak in the explanted chick blastoderm marked with carbon particles. J. Exptl. Zool., **103**:259–304.

_____, 1947. Regression and shortening of the primitive streak in the explanted chick blastoderm. J. Exptl. Zool., **104**:69–100.

_____, 1955. Analysis of the organizer center in the early chick embryo. I. Localization of prospective notochord and somite cells. J. Exptl. Zool., **128**:121–164.

_____, 1957. Analysis of the organizer center in the early chick embryo. II. Studies of the mechanics of notochord elongation and somite formation. J. Exptl. Zool., 134:577–612.

———— and Haas, H., 1960. Integrative mechanisms in development of the early chick blastoderm. I. Regulative potentiality of separated parts. J. Exptl. Zool., 145:97–137.

———— and Haas, H., 1961. Integrative mechanisms in development of the early chick blastoderm. II. Role of morphogenetic movements and regenerative growth in synthetic and topographically disarranged blastoderms. J. Exptl. Zool., 147:57–93.

Waddington, C. H., 1933a. Induction by the endoderm in birds. Arch. Entw.-mech. Organ., 128:502–521.

————, 1933b. Induction by the primitive streak and its derivatives in the chick. J. Exptl. Biol., 10:38–46.

————, 1933c. Induction by coagulated organisers in the chick embryo. Nature, 131:275.

————, 1934. Experiments on embryonic induction. I. The competence of the extra-embryonic ectoderm in the chick. II. Experiments on coagulated organisers in the chick. III. A note on inductions by chick primitive streak transplanted to the rabbit embryo. J. Exptl. Biol., 11:211–227.

———— and Cohen, A., 1936. Experiments on the development of the head of the chick embryo. J. Exptl. Biol., 13:219–236.

———— and Schmidt, G. A., 1933. Induction by heteroplastic grafts of the primitive streak in birds. Arch. Entw.-mech. Organ., 128:521–563.

Willier, B. H., 1954. Phases in embryonic development. J. Cell. Comp. Physiol., 43(suppl. 1):307–318.

———— and Rawles, M. E., 1931. The relation of Hensen's node to the differentiating capacity of whole chick blastoderms as studied in chorioallantoic grafts. J. Exptl. Zool., 59:429–465.

———— and Rawles, M. E., 1935. Organ-forming areas in the early chick blastoderm. Proc. Soc. Exptl. Biol. Med., 32:1293–1296.

———— and Yuh, E. C., 1928. The problem of sex differentiation in the chick embryo with reference to the effects of gonad and non-gonad grafts. J. Exptl. Zool., 52:65–125.

Young, R. T., 1919. Some abnormalities of notochord and neural tube in the chick. Quart. J. Univ. North Dakota, 9:231–234.

Zacks, S. I., 1954. Esterases in the early chick embryo. Anat. Record, 118:509–537.

V. Extra-embryonic Membranes

Adamstone, F. B., 1948. Experiments on the development of the amnion in the chick. J. Morphol., 83:359–371.

Bautzmann, H., Schmidt, W., and Lemburg, P., 1960. Experimental electron- and light-microscopical studies on the function of the amnion-apparatus of the chick, cat and man. Anat. Anz., **108**:305–315.

————, and Schröder, R., 1953. Studien zur funktionellen Histologie und Histogenese des Amnions beim Hühnchen und beim Menschen. Zeitschr. f. Anat. u. Entwicklungsges., **117**:166–214.

Danchakoff, V., 1917. The position of the respiratory vascular net in the allantois of the chick. Am. J. Anat., **21**:407–420.

Duval, M., 1884. Études histologiques et morphologiques sur les annexes des embryons d'oiseaux. J. de l'Anat. et de la Physiol., **20**:201–241.

Ebert, J. D., 1963. The labile chorioallantoic membrane: Changes effected by viruses and inhibitors. Am. Zool., **3**:235–243.

Gruenwald, P., 1941. Normal and abnormal detachment of body and gut from the blastoderm in the chick embryo, with remarks on the early development of the allantois. J. Morphol., **69**:83–125.

Hanan, E. B., 1927. Absorption of vital dyes by the fetal membranes of the chick. I. Vital staining of the chick embryo by injections of trypan blue into the air chamber. Am. J. Anat., **38**:423–450.

Kugler, O. E., 1945. Acid-soluble phosphorus in the amniotic and allantoic fluids of the developing chick. J. Cellular Comp. Physiol., **25**:155–160.

Latta, J. S., and Busby, L. F., 1929. The reaction of the chick embryo and its membranes to trypan blue. Am. J. Anat., **44**:171–198.

Lillie, F. R., 1903. Experimental studies on the development of the organs in the embryo of the fowl (Gallus domesticus). I. Experiments on the amnion and the production of anamniote embryos of the chick. Biol. Bull., **5**:92–124.

Pierce, M. E., 1933. The amnion of the chick as an independent effector. J. Exptl. Zool., **65**:443–473.

Popoff, D., 1894. Die Dottersack-gefässe des Huhnes. C. W. Kreidel, Wiesbaden, 43 pp.

Romanoff, A. L., and Romanoff, A. J., 1933. Gross assimilation of yolk and albumen in the development of the egg of Gallus domesticus. Anat. Record, **55**:271–278.

Wislocki, G. B., 1921. Note on the behavior of trypan blue injected into the developing egg of the hen. Anat. Record, **22**:267–274.

Zwilling, E., 1946. Regulation in the chick allantois. J. Exptl. Zool., **101**:445–453.

VI.   Nervous System and Sense Organs

Adelmann, H. B., 1927. The development of the eye muscles of the chick. J. Morphol. Physiol., **44**:29–87.

Alexander, L. E., 1937. An experimental study of the role of optic cup and overlying ectoderm in lens formation in the chick embryo. J. Exptl. Zool., **75**:41–73.

Andrew, A., 1964. The origin of intramural ganglia. I. The early arrival of precursor cells in the presumptive gut of chick embryos. J. Anat., London, **98**:421–428.

Barron, D. H., 1948. Some effects of amputation of the chick wing bud on the early differentiation of the motor neuroblasts in the associated segments of the spinal cord. J. Comp. Neurol., **88**:93–127.

Bergquist, H., 1959. Experiments on the 'overgrowth' phenomenon in the brain of chick embryos. J. Embryol. Exptl. Morphol., **7**:122–127.

Bernd, A. H., 1905. Die Entwickelung des Pecten im Auge des Hühnchens aus den Blätten des Augenblase. Diss. med. Bonn.

Birge, W. J., 1959. Spontaneous alar plate hyperplasia in the chick embryo. Anat. Record, **135**:135–140.

Brizzee, K. R., 1949. Histogenesis of the supporting tissue in the spinal and the sympathetic trunk ganglia in the chick, J. Comp. Neurol, **91**:129–146.

Bueker, E. D., 1947. Limb ablation experiments on the embryonic chick and its effect as observed on the mature nervous system. Anat. Record, **97**:157–174.

Cajal, S. Ramón y, 1889 and 1890. *See* Ramón y Cajal, S.

Carpenter, F. W., 1906. The development of the oculomotor nerve, the ciliary ganglion, and the abducent nerve in the chick. Bull. Museum Comp. Zool., Harvard Coll., **48**:139–229.

Clarke, L. F., 1936. Regional differences in eye-forming capacity of the early chick blastoderm as studied in chorio-allantoic grafts. Physiol. Zool., **9**:102–128.

Clarke, W. M., and Fowler, I., 1960. The inhibition of lens-inducing capacity of the optic vesicle with adult lens antisera. Develop. Biol., **2**:155–172.

Clearwaters, K. P., 1954. Regeneration of the spinal cord of the chick. J. Comp. Neurol., **101**:317–329.

Cohn, F., 1903. Zur Entwickelungsgeschichte des Geruchsorgans des Hühnchens. Arch. Entw.-mech. Organ., **61**:133-150.

Corliss, C. E., and Robertson, G. G., 1963. The pattern of mitotic density in the early chick neural epithelium. J. Exptl. Zool., **153**:125–140.

Coulombre, A. J., 1955. Correlations of structural and biochemical changes in the developing retina of the chick. Am. J. Anat., **96**:153–189.

————, 1956. The role of intraocular pressure in the development of the chick eye. I. Control of eye size. J. Exptl. Zool., **133**:211–226.

———— and Coulombre, J. L., 1958. Corneal development. I. Corneal transparency. J. Cellular Comp. Physiol., **51**:1–11.

————, Steinberg, S. N., and Coulombre, J. L., 1963. The role of intraocular pressure in the development of the chick eye. V. Pigmented epithelium, Invest. Ophthal., **2**:83–89.

Coulombre, J. L., and Coulombre, A. J., 1963. Lens development: Fiber elongation and lens orientation. Science, **142**:1489–1490.

Cowdry, E. B., 1914. The development of the cytoplasmic constituents of the nerve cells of the chick. Am. J. Anat., **15**:389–430.

Danchakoff, V., 1926. Lens ectoderm and optic vesicles in allantois grafts. Carnegie

Cont. to Embryol., **18:**63–78.

———— and Agassiz, A., 1924. Growth and development of the neural plate in the allantois. J. Comp. Neurol., **37:**397–438.

Dexter, F., 1902. The development of the paraphysis in the common fowl. Am. J. Anat., **2:**13–24.

Dorris, F., 1938. Differentiation of the chick eye in vitro. J. Exptl. Zool., **78:**385–415.

————, 1941. The behavior of chick neural crest in grafts to the chorio-allantoic membrane. J. Exptl. Zool., **86:**205–223.

Doyle, W. L., and Watterson, R. L., 1949. The accumulation of glycogen in the "glycogen body" of the nerve cord of the developing chick. J. Morphol., **85:**391–404.

Eckardt, L. B., and Elliott, R., 1935. The development of the motor nuclei of the hindbrain of the chick, Gallus domesticus. J. Comp. Neurol., **61:**83–99.

Elchlepp, J. G., 1956. Development of the chick eye: Relation of ground substance change to organ growth. Anat. Record, **126:**425–432.

Feeney, J. F., Jr., and Watterson, R. L., 1946. The development of the vascular pattern within the walls of central nervous system of the chick embryo. J. Morphol., **78:**231–303.

Fowler, I., 1953. Responses of the chick neural tube in mechanically produced spina bifida. J. Exptl. Zool., **123:**115–152.

Gayer, K., 1942. A study of coloboma and other abnormalities in transplants of eye primordia from normal and Creeper chick embryos. J. Exptl. Zool., **89:**103–133.

Gruenwald, P., 1944. Studies on developmental pathology. I. The morphogenesis of a hereditary type of microphthalmia in chick embryos. Anat. Record, **88:**67–81.

Hamburger, V., 1939. The development and innervation of transplanted limb primordia of chick embryos. J. Exptl. Zool., **80:**347–389.

————, 1946. Isolation of the brachial segments of the spinal cord of the chick embryo by means of tantalum foil blocks. J. Exptl. Zool., **103:**113–142.

————, 1948. The mitotic patterns in the spinal cord of the chick embryo and their relation to histogenetic processes. J. Comp. Neurol. **88:**221–283.

————, 1961. Experimental analysis of the dual origin of the trigeminal ganglion in the chick embryo. J. Exptl. Zool., **148:**91–123.

———— and Keefe, E. L., 1944. The effects of peripheral factors on the proliferation and differentiation in the spinal cord of chick embryos. J. Exptl. Zool., **96:**223–242.

———— and Levi-Montalcini, R., 1949. Proliferation, differentiation and degeneration in the spinal ganglia of the chick embryo under normal and experimental conditions. J. Exptl. Zool., **111:**457–501.

Hammond, W. S., 1949. Formation of the sympathetic nervous system in the trunk of the chick embryo following removal of the neural tube. J. Comp. Neurol., 91:67–85.

⸻ and Yntema, C. L., 1947. Depletions in the thoraco-lumbar sympathetic system following removal of neural crest in the chick. J. Comp. Neurol. 86:237–265.

⸻ and Yntema, C. L., 1958. Origin of ciliary ganglia in the chick. J. Comp. Neurol., 110:367–390.

Harrison, J. R., 1954. Growth and differentiation of the embryonic chick eye in vitro. I. On yolk-albumen fractions obtained with centrifugation and heat. J. Exptl. Zool., 127:493–510.

Harrison, R. G., 1908. Embryonic transplantation and the development of the nervous system. Anat. Record, 2:385–410.

Held, H., 1909. Die Entwicklung des Nervengewebes bei den Wirbeltieren. J. A. Barth, Leipzig. ix + 378 pp.

Herrick, C. J., 1909. The criteria of homology in the peripheral nervous system. J. Comp. Neurol., 19:203–209.

⸻, 1925. Morphogenetic factors in the differentiation of the nervous system. Physiol. Rev., 5:112–130.

Hill, C., 1900. Developmental history of the primary segments of the vertebrate head. Zoöl. Jahrb., Anat. Abt., 13:393–446.

Hobson, L. B., 1941. On the ultrastructure of the neural plate and tube of the early chick embryo, with notes on the effects of dehydration. J. Exptl. Zool., 88: 107–134.

Holmdahl, D. E., 1928. Die Entstehung und weitere Entwicklung der Neuralleiste (Ganglienleiste) bei Vögeln und Säugetieren. Zeitschr. f. mikr.-anat. Forsch., 14:99–298.

Jones, D. S., 1937. The origin of the sympathetic trunks in the chick embryo. Anat. Record, 70:45–65.

⸻, 1941. Further studies on the origin of sympathetic ganglia in the chick embryo. Anat. Record, 79:7-15.

⸻, 1942. The origin of the vagi and the parasympathetic ganglion cells of the viscera of the chick. Anat. Record, 82:185–197.

⸻, 1945. The origin of the ciliary ganglia in the chick embryo. Anat. Record, 92:441–447.

Joy, E. A., 1939. Intra-coelomic grafts of the eye primordium of the chick. Anat. Record, 74:461–485.

Källén, B., 1953. On the significance of the neuromeres and similar structures in vertebrate embryos. J. Embryol. Exptl. Morphol., 1:387–392.

Kingsbury, B. F., 1922. The fundamental plan of the vertebrate brain. J. Comp. Neurol., 34:461–492.

⸻, 1926. Branchiomerism and the theory of head segmentation. J. Morphol. Physiol., 42:83–109.

⸻ and Adelmann, H. B., 1924. The morphological plan of the head. Quart.

J. Microscop. Sci., **68**:239–285.

Krabbe, K. H., 1955. Development of the pineal organ and a rudimentary parietal eye in some birds. J. Comp. Neurol., **103**:139–149.

Kuhlenbeck, H., 1937. The ontogenetic development of the diencephalic centers in a bird's brain (chick) and comparison with the reptilian and mammalian diencephalon. J. Comp. Neurol., **66**:23–75.

Kunkel, B. W., 1928. On an abnormal optic vesicle in a chick embryo. Anat. Record, **37**:381–388.

Kuntz, A., 1922. Experimental studies on the histogenesis of the sympathetic nervous system. J. Comp. Neurol., **34**:1–36.

———, 1926. The rôle of cells of medullary origin in the development of the sympathetic trunks. J. Comp. Neurol., **40**:389–408.

——— and Batson, O. V., 1920. Experimental observations on the histogenesis of the sympathetic trunks in the chick. J. Comp. Neurol. **32**:335–346.

Langman, J., 1959. The first appearance of specific antigens during the induction of the lens. J. Embryol. Exptl. Morphol., **7**:193–202 and 264–276.

Levi-Montalcini, R., 1949. The development of the acoustico-vestibular centers in the chick embryo in the absence of the afferent root fibers and of descending fiber tracts. J. Comp. Neurol., **91**:209–241.

———, 1950. The origin and development of the visceral system in the spinal cord of the chick embryo. J. Morphol., **86**:253–283.

Lewis, W. H., 1903. Wandering pigmented cells arising from the epithelium of the optic cup, with observations on the origin of the m. sphincter pupillae in the chick. Am. J. Anat., **2**:405–416.

Locy, W. A., 1895. Contributions to the structure and development of the vertebrate head. J. Morphol., **11**:497–594.

———, 1897. Accessory optic vesicles in the chick embryo. Anat. Anz., **14**:113–124.

McKeehan, M. S., 1954. A quantitative study of self-differentiation of transplanted lens primordia in the chick. J. Exptl. Zool., **126**:157–176.

Mall, F., 1888. Development of the Eustachian tube, middle ear, tympanic membrane and meatus of the chick. Stud. Johns Hopkins Univ., **4**:185–192.

Mangold, O., 1931. Das Determinationsproblem. 3. Das Wirbeltierauge in der Entwicklung und Regeneration. Ergeb. Biol., **7**:193–402.

Marsh, G., and Beams, H. W., 1946. In vitro control of growing chick nerve fibers by applied electric currents. J. Cellular Comp. Physiol., **27**:139–157.

Martin, A., and Langman, J., 1965. The development of the spinal cord examined by autoradiography. J. Embryol. Exptl. Morphol., **14**:25–35.

Menkes, B., Alexandru, C., and Tudose, O., 1967. Investigation on pre- and postmitotic movement of the neuroblasts, in the embryonic neural tube. Rev. Roumaine Embryol. Cytol. **4**:29–33.

Meyer, D. B., and O'Rahilly, R., 1959. The development of the cornea in the chick, J. Embryol. Exptl. Morphol., **7**:303–315.

Nakai, J., 1956. Dissociated dorsal root ganglia in tissue culture. Am. J. Anat., **99**: 81–130.

Nawar, G., 1956. Experimental analysis of the origin of the autonomic ganglia in the chick embryo. Am. J. Anat., **99**:473–506.

Neal, H. V., 1918. The history of the eye muscles. J. Morphol., **30**:433–453.

O'Rahilly, R., 1962. The development of the sclera and the choroid in staged chick embryos. Acta Anat., **48**:335–346.

———and Meyer, D. B., 1960. The periodic acid-Schiff reaction in the cornea of the developing chick. Z. Anat. Entwicklg., **121**:351–368.

Orr, D. W., and Windle, W. F., 1934. The development of behavior in chick embryos: The appearance of somatie movements. J. Comp. Neurol., **60**:271–285.

Ramón y Cajal, S., 1889. Sur la morphologie et les connexions des elements de la retine des oiseaux. Anat. Anz., **4**:111–121.

———, 1890. Sur l'origine et les ramifications des fibres nerveuses de la moelle embryonnaire. Anat. Anz., **5**:85–95; 111–119.

Raybuck, H. E., 1956. Experimental data on the histogenesis of ganglion cells in the sympathetic trunk of the chick. Anat. Record, **124**:603–618.

Retzius, G., 1881–1884. Das Gehörorgan der Wirbelthiere. II. Das Gehörorgan der Reptilien, der Vögel und der Säugethiere. Stockholm, viii + 368 pp.

Rhines, R., and Windle, W. F., 1944. An experimental study of factors influencing the course of nerve fibers in the embryonic central nervous system. Anat. Record, **90**:267–294.

Rogers, K. T., 1957. Early development of the optic nerve in the chick. Anat. Record, **127**:97–107.

Rothig, P., and Brugsch, T., 1902. Die Entwicklung des Labyrinthes beim Huhn. Arch. f. mikr. Anat., **59**:354–388.

Sauer, F. C., 1935a. Mitosis in the neural tube. J. Comp Neurol., **62**:377–405.

———, 1935b. The cellular structure of the neural tube. J. Comp. Neurol., **63**: 13–23.

Schumacher, S., 1927. Über die sogenannte Vervielfachung des Medullarrohres (bzw. des Canalis centralis) bei Embryonen. Zeitschr. f. mikr.-anat. Forsch., **10**:75–109.

Shen, S., Greenfield, P., and Boell, E. J., 1956. Localization of acetylcholinesterase in chick retina during histogenesis. J. Comp. Neurol., **106**:433–461.

Sjodin, R. A., 1957. The behavior of brain and retinal tissue in mortality of the early chick embryo. Anat. Record, **127**:591–610.

Spratt, N. T., Jr., 1940. An in vitro analysis of the organization of the eye-forming area in the early chick blastoderm. J. Exptl. Zool., **85**:171–209.

Stockard, C. R., 1914. The artificial production of eye abnormalities in the chicken embryo. Anat. Record, **8**:33–42.

Streeter, G. L., 1933. The status of metamerism in the central nervous system of chick embryos. J. Comp. Neurol., **57**:455–475.

Van Campenhout, E., 1931. Le développement du Système Nerveux Sympathique chez le Poulet. Extrait des Arch. Biol. **42**:479–505.

————, 1933. The innervation of the digestive tract in the 6-day chick embryo. Anat. Record, 56:11–118.

Watterson, R. L., 1949. Development of the glycogen body of the chick spinal cord. I. Normal morphogenesis, vasculogenesis and anatomical relationships. J. Morphol., 85:337–389.

————, 1954. Development of the glycogen body of the chick spinal cord. IV. Effects of mechanical manipulation of the roof plate at the lumbosacral level. J. Exptl. Zool., 125:285–330.

————and Fowler, I., 1953. Regulative development in lateral halves of chick neural tubes. Anat. Record, 117:773–803.

————, Fowler, I., and Fowler, B. J., 1954. The role of the neural tube and notochord in development of the axial skeleton of the chick. Am. J. Anat., 95:337–399.

————, Goodheart, C. R., and Lindberg, G., 1955. The influence of adjacent structures upon the shape of the neural tube and neural plate of chick embryos. Anat. Record, 122:539–559.

————and Spiroff, B. E. N., 1949. Development of the glycogen body of the chick spinal cord. II. Effects of unilateral and bilateral leg-bud extirpation. Physiol. Rev., 22:318–337.

Weiss, P. A., 1934. In vitro experiments on the factors determining the course of the outgrowing nerve fiber. J. Exptl. Zool., 68:393–448.

————, 1945. Experiments on cell and axon orientation in vitro: the role of colloidal exudates in tissue organization. J. Exptl. Zool., 100:353–386.

————and Wang, H., 1936. Neurofibrils in living ganglion cells of the chick, cultivated in vitro. Anat. Record, 67:105–117.

Weysse, A. W., and Burgess, W. S., 1906. Histogenesis of the retina. Am. Naturalist, 40:611–638.

Williams, R. G., 1937. The development of vascularity in the hindbrain of the chick. J. Comp. Neurol., 66:77–101.

Windle, W. F., and Austin, M. F., 1936. Neurofibrillar development in the central nervous system of chick embryos up to 5 days' incubation. J. Comp. Neurol., 63:431–463.

————and Orr, D. W., 1934. The development of behavior in chick embryos: spinal cord structure correlated with early somatic motility. J. Comp. Neurol., 60: 287–307.

Yntema, C. L., and Hammond, W. S., 1955. Experiments on the origin and development of the sacral autonomic nerves in the chick embryo. J. Exptl. Zool., 129: 375–413.

VII. Circulatory System

Bakst, H. J., and Chafee, F. H., 1928. The origin of the definitive subclavian artery in the chick embryo. Anat. Record, 38:129–140.

Barry, A., 1942. The intrinsic pulsation rates of fragments of the embryonic chick heart. J. Exptl. Zool., **91**:119–130.

————, 1948. The functional significance of the cardiac jelly in the tubular heart of the chick embryo. Anat. Record, **102**:289–298.

Bogue, J. Y., 1933. The electrocardiogram of the developing chick. J. Exptl. Biol., **10**:286–292.

Boucek, R. J., Murphy, W. P., and Paff, G. H., 1959. Electrical and mechanical properties of chick embryo heart chambers. Circulation Res., **7**:787–793.

Bremer, J. L., 1926. The influence of nerves on the position of the coeliac artery in the chick. Anat. Record, **33**:299–310.

————, 1928a. Experiments on the aortic arches in the chick. Anat. Record, **37**:225–254.

————, 1928b. I. An interpretation of the development of the heart. Am. J. Anat., **42**:307–337.

————, 1932. Circulatory disturbances in operated chick embryos: Reversal of heart beat. Anat. Record, **51**:275–284.

Buell, C. E., 1922. Origin of the pulmonary vessels in the chick. Carnegie Cont. to Embryol., **14**:11–26.

Cavanaugh, M. W., 1955. Pulsation, migration and division in dissociated chick embryo heart cells in vitro. J. Exptl. Zool., **128**:573–590.

———— and Cavanaugh, D. J., 1957. Studies on the pharmacology of tissue cultures. I. The action of quinidine on cultures of dissociated chick embryo heart cells. Arch. Int. Pharmacodyn. Therap., **110**:43–55.

Chang, C., 1931. On the origin of the pulmonary vein. Anat. Record, **50**:1–8.

Chapman, W. B., 1918. The effect of the heart-beat upon the development of the vascular system in the chick. Am. J. Anat., **23**:175–203.

Clark, E. L., 1915. Observations on the lymph flow and the associated morphological changes in the early superficial lymphatics of chick embryos. Am. J. Anat., **18**:399–440.

Clark, E. R., and Clark, E. L., 1920. On the origin and early development of the lymphatic system of the chick. Cont. to Embryol., Carnegie Inst. Wash., No. 45, Mall Memorial Vol., pp. 442–481.

Cohn, A. E., 1928. Physiological ontogeny. A. Chicken embryos. XIII. The temperature characteristic for the contraction rate of the whole heart. J. Gen. Physiol., **11**:369–375.

Congdon, E. D., 1918. The embryonic structure of avian heart muscle, with some considerations regarding its earliest contraction. Anat. Record, **15**:135–150.

———— and Wang, H. W., 1926. The mechanical processes concerned in the formation of the differing types of aortic arches of the chick and the pig and in the divergent early development of their pulmonary arches. Am. J. Anat., **37**:499–520.

Danchakoff, V., 1916. Origin of the blood cells. Development of haematopoietic organs and regeneration of the blood cells from the standpoint of the monophyletic school. Anat. Record, **10**:397–413.

———, 1917. The position of the respiratory vascular net in the allantois of the chick. Am. J. Anat., **21**:407–420.

———, 1918. Equivalence of different hematopoietic anlages (by method of stimulation of their stemcells). II. Grafts of adult spleen on the allantois and response of the allantoic tissues. Am. J. Anat., **24**:127–189.

Davis, C. L., 1924. The cardiac jelly of the chick embryo. Anat. Record, **27**:201–202.

DeHaan, R. L., 1959. Cardia bifida and the development of pacemaker function of the early chick heart. Develop. Biol., **1**:586–602.

———, 1963a. Organization of the cardiogenic plate in the early chick embryo. Acta Embryol. Morphol. Experim., **6**:26–38.

———, 1963b. Regional organization of prepacemaker cells in the cardiac primordia of the early chick embryo. J. Embryol. Exptl. Morphol., **11**:65–76.

———, 1964. Cell interactions and oriented movements during development. J. Exptl. Zool., **157**: 127–138.

———, 1967. Development of form in the embryonic heart. Circulation, **35**:821–833.

Doan, C. A., Cunningham, R. S., and Sabin, F. R., 1925. Experimental studies on the origin and maturation of avian and mammalian red blood-cells. Carnegie Cont. to Embryol., **16**:163–226.

Douarin, G. L., 1960. Les malformations cardiaques obtenues par l'irradiation aux rayon X du coeur de l'embryon de poulet. J. Embryol. Exptl. Morphol., **8**:130–138.

———, 1963. Étude de l'action des rayons X sur la physiologie et l'organogenèse du coeur embryonnaire chez le poulet. Bull. Biol. France Belgique, **97**:643–759.

———, 1968. Quelques aspects de l'action des rayons X sur la morphogenese et la physiologie d'organes embryonnaires. Ann. Fac. Sci. Univ. Clemont, **39**: 5–31.

Evans, H. M., 1909a. On the development of the aortae, cardinal and umbilical veins and other blood-vessels of vertebrate embryos from capillaries. Anat. Record, **3**:498–518.

———, 1909b. On the earliest blood-vessels in the anterior limb buds of birds and their relation to the primary subclavian artery. Am. J. Anat., **9**:281–321.

Fingl, E., Woodbury, L. A., and Hecht, H. H., 1952. Effects of innervation and drugs upon direct membrane potentials of embryonic chick myocardium. J. Pharmacol. Exptl. Therapeutics, **104**:103–114.

Fleming, R. E., 1926. The origin of the vertebral and external carotid arteries in birds. Anat. Record, **33**:183–199.

Graper, L., 1907. Untersuchungen über die Hertzbildung der Vögel. Arch. Entw.-mech. Organ., **24**:375–410.

Hahn, H., 1909. Experimentelle Studien über die Entstehung des Blutes und der ersten Gefässe beim Hühnchen. Arch. Entw.-mech. Organ., **27**:337–433.

Hibbs, R. G., 1956. Electron microscopy of developing cardiac muscle in chick embryos. Am. J. Anat., **99**:17–51.

Hochstetter, F., 1906. Die Entwickelung des Blutgefässystems. Hertwig's Handbuch der vergleichenden und experimentellen Entwickelungslehre der Wirbeltiere. III. Pt. 2, 21–166.

Hoff, E. C., Kramer, T. C., DuBois, D., and Patten, B. M., 1939. The development of the electrocardiogram of the embryonic heart. Am. Heart J., **17**:470–488.

Hughes, A. F. W., 1934. On the development of the blood vessels in the head of the chick. Phil. Trans. Roy. Soc. London, **224**:75–161.

Hyman, L. H., 1927. The metabolic gradients of vertebrate embryos. IV. The heart. Biol. Bull., **52**:39–50.

Jaffe, O. C., 1965. Hemodynamic factors in the development of the chick embryo heart. Anat. Record, **151**:69–76.

———, 1967. The development of the arterial outflow tract in the chick embryo heart. Anat. Record, **158**:35–42.

Johnstone, P. N., 1924. Studies on the physiological anatomy of the embryonic heart. I. The demonstration of complete heart block in chick embryos during the second, third and fourth days of incubation. Johns Hopkins Hosp. Bull., **35**: 87–90.

———, 1925. Studies on the physiological anatomy of the embryonic heart. II. An inquiry into the development of irritability to electrical stimulation. Johns Hopkins Hosp. Bull., **36**:299–311.

Jones, D. S., 1958. Effects of acetylcholine and adrenalin on the experimentally uninnervated heart of the chick embryo. Anat. Record, **130**:253–259.

Lewis, M. R., 1919. The development of cross striation in the heart muscle of the chick embryo. Bull. Johns Hopkins Hosp., **30**:176–181.

Lewis, W. H., 1926. Cultivation of embryonic heart-muscle. Carnegie Cont. to Embryol., **18**:1–21.

Lieberman, M. and Paes de Carvalho, A., 1965a. The electro-physiological organization of the embryonic chick heart. J. Gen. Physiol., **49**:351–363.

——— and Paes de Carvalho, A., 1965b. The spread of excitation in the embryonic chick heart. J. Gen. Physiol., **49**:365–379.

Locy, W. A., 1906. The fifth and sixth aortic arches in chick embryos with comments on the condition of the same vessels in other vertebrates. Anat. Anz., **29**:287–300.

Menkes, B., Alexandru, C., Pavkov, A., and Mircov, O., 1965. Researches on the formation and the elastic structure of the aorto-pulmonary septum in the chick embryo. Rev. Roumaine Embryol. Cytol., **2**:79–91.

Miller, A. M., 1903. The development of the postcaval vein in birds. Am. J. Anat., **2**:283–298.

———, 1913. Histogenesis and morphogenesis of the thoracic duct in the chick; development of the blood cells and their passage into the blood stream via thoracic duct. Am. J. Anat., **15**:131–197.

——— and McWhorter, J. E., 1914. Experiments on the development of blood vessels in the area pellucida and embryonic body of the chick. Anat. Record 8:203–227.

Murray, H. A., Jr., 1926. Physiological ontogeny. A. Chicken embryos. X. The temperature characteristic for the contraction rate of isolated fragments of embryonic heart muscle. J. Gen. Physiol., 9:781–788.

Nace, G. W., 1953. Serological studies of the blood of the developing chick embryo. J. Exptl. Zool., 122:423–448.

Olivo, O., 1925. Sull'inizio della funzione contracttile del cuore e dei miotomi dell' embrione di pollo in rapporto alla loro differenziazione morfologica e strutturale. Arch. Exptl. Zellforsch., 1:427–500.

Paff, G. H., 1935. Conclusive evidence for sino-atrial dominance in isolated 48-hour embryonic chick hearts cultivated *in vitro*. Anat. Record, 63:203–212.

————, 1936. Transplantation of sino-atrium to conus in the embryonic heart *in vitro*. Am. J. Physiol., 117:313–317.

———— and Boucek, R. J., 1962. Simultaneous electrocardiograms and myograms of the isolated atrium, ventricle and conus of the embryonic chick heart. Anat. Record, 142:73–80.

Patten, B. M., 1922. The formation of the cardiac loop in the chick. Am. J. Anat., 30:373–397.

————, 1925. The interatrial septum of the chick heart. Anat. Record, 30:53–60.

————, 1939. Microcinematographic and electrocardiographic studies of the first heart beats and the beginning of the circulation in living embryos. Proc. Inst. Med. Chicago, 12:366–380.

————, 1949. Initiation and early changes in the character of the heart beat in vertebrate embryos. Physiol. Rev., 29:31–47.

———— and Kramer, T. C., 1932. A moving-picture apparatus for microscopic work. Anat. Record, 52:169–189.

———— and Kramer, T. C., 1933. The initiation of contraction in the embryonic chick heart. Am. J. Anat., 53:349–375.

————, Kramer, T. C., and Barry, A., 1948. Valvular action in the embryonic chick heart by localized apposition of endocardial masses. Anat. Record, 102:299–311.

Patterson, J. T., 1909. An experimental study on the development of the vascular area of the chick blastoderm. Biol. Bull., 16:83–90.

Pohlman, A. G., 1920. A consideration of the branchial arcades in chick based on the anomalous persistence of the fourth left arch in a sixteen-day stage. Anat. Record, 18:159–166.

Popoff, D., 1894. Die Dottersack-gefässe des Huhnes. C. W. Kreidel, Wiesbaden. 43 pp.

Reagan, F. P., 1917. Experimental studies on the origin of vascular endothelium and of erythrocytes. Am. J. Anat., 21:39–175.

Rényi, G. S. de, and Hogue, M. J., 1938. Studies on cardiac muscle cells, from chick embryos, grown in tissue culture. Anat. Record, 70:441–449.

Romanoff, A. L., 1944. The heart beat of avian embryos. Anat. Record, **89**:313–316.

Rosenquist, G. C., 1966. A radioautographic study of labeled grafts in the chick blastoderm. Development from primitive-streak stages to stage 12. Carnegie Contrib. to Embryol., **38**:71–110.

———— and DeHaan, R. L., 1966. Migration of precardiac cells in the chick embryo: A radioautographic study. Carnegie Contrib. to Embryol., **38**:111–121.

Sabin, F. R., 1917. Origin and development of the primitive vessels of the chick and the pig. Carnegie Cont. to Embryol., **6**:61–124.

————, 1920. Studies on the origin of blood-vessels and of red blood-corpuscles as seen in the living blastoderm of chicks during the second day of incubation. Carnegie Cont. to Embryol., Mall Memorial Vol., pp. 213–262.

————, 1921. Studies on blood. The vitally stainable granules as a specific criterion for erythroblasts and the differentiation of the three strains of the white blood cells as seen in the living chick's yolk sac. Johns Hopkins Hosp. Bull., **32**:314–321.

————, 1922. Direct growth of veins by sprouting. Carnegie Cont. to Embryol., No. 65, **14**:1–10.

Sangvichien, S., 1952. Experimental reversal of the heart beat in chick embryos. Anat. Record, **112**:529–538.

Schlater, G., 1907. Histologische Untersuchungen über das Muskelgewebe. Die Myofibrillen des embryonalen Hühnerherzens. Arch. f. mikr. Anat., **69**:100–116.

Shipley, P. G., 1916. The development of erythrocytes from hemoglobin-free cells and the differentiation of heart muscle fibers in tissue cultivated in plasma. Anat. Record, **10**:347–353.

Sippel, T. O., 1955. Properties and development of cholinesterase in the hearts of certain vertebrates. J. Exptl. Zool., **128**:165–184.

Squier, T. L., 1916. On the development of the pulmonary circulation in the chick. Anat. Record, **10**:425–438.

Stockard, C. R., 1915. An experimental analysis of the origin of blood and vascular endothelium. Memoirs Wistar Inst., No. 7, 174 pp.

Stalsberg, H., and DeHaan, R. L., 1969. The precardiac areas and formation of the tubular heart in the chick embryo. Develop. Biol., **19**:128–159.

Sugiyama, S., 1926. Origin of thrombocytes and the different types of blood-cells as seen in the living chick blastoderm. Carnegie Cont. to Embryol., **18**:121–147.

Tonge, M., 1869. On the development of the semilunar valves of the aorta and pulmonary artery of the chick. Phil. Trans. Roy. Soc. London, **159**:387–411.

Van Mierop, L. H. S., 1967. Location of pacemaker in chick embryo heart at the time of initiation of heartbeat. Am. J. Physiol., **212**:407–415.

Wolff, E., 1933. Realisation experimentale de monstres a coeur double. Compt. Rend. Soc. Biol., **112**:1090–1092.

———— and Stephan, F., 1948. Analyse experimentale de la determination des gros vaisseaux extra-embryonnaires chez l'oiseau. Compt. Rend. Soc. Biol., **142:** 1018–1020.

VIII. Digestive and Respiratory Systems and Body Cavities

Boyden, E. A., 1918. Vestigial gill filaments in chick embryos with a note on similar structures in reptiles. Am. J. Anat., 23:205–235.

———, 1924. An experimental study of the development of the avian cloaca, with special reference to a mechanical factor in the growth of the allantois. J. Exptl. Zool., **40:**437–471.

———See also references under The Urogenital System.

Brouha, M., 1898. Recherches sur le développement du foie, du pancréas, de la cloison mesentérique et des cavités hépato-entériques chez les oiseaux. J. de l'Anat. et de la Physiol., **34:**305–363.

Butler, G. W., 1899. On the subdivisions of the body-cavity in lizards, crocodiles, and birds. Proc. Zool. Soc. London, pp. 452–474.

Dalton, A. J., 1937. The functional differentiation of the hepatic cells of the chick embryo. Anat. Record, **68:**393–405.

Dudley, J., 1942. The development of the ultimobranchial body of the fowl, Gallus domesticus. Am. J. Anat., **71:**65–97.

Hildebrandt, W., 1903. Die erste Leberentwickelung beim Vogel. Anat. Hefte, **20:** 73–120.

Hunt, T. E., 1937. The development of gut and its derivatives from the mesectoderm and mesentoderm of early chick blastoderms. Anat. Record, **68:**349–369.

Ivey, W. D., and Edgar, S. A., 1952. The histogenesis of the esophagus and crop of the chicken, turkey, guinea fowl and pigeon, with special reference to ciliated epithelium. Anat. Record, **114:**189–212.

Kingsbury, J. W., Alexanderson, M., and Kornstein, E. S., 1956. The development of the liver in the chick. Anat. Record, **124:**165–188.

Locy, W. A., and Larsell, O., 1916. The embryology of the bird's lung based on observations of the domestic fowl. Am. J. Anat., **19:**447–504; **20:**1–44.

Lynch, R. S., 1921. The cultivation of liver cells of the chick embryo in vitro. Am. J. Anat., **29:**281–311.

Mall, F. P., 1891. Development of the lesser peritoneal cavity in birds and mammals. J. Morphol., **5:**165–179.

Minot, C. S., 1900a. On the solid stage of the large intestine in the chick, with a note on the ganglion coli. J. Boston Soc. Med. Sci., **4:**153–164.

———, 1900b. On a hitherto unrecognized form of blood-circulation without capillaries in the organs of vertebrata. Proc. Boston Soc. of Nat. Hist., **29:**185–215.

Moog, F., 1950. The functional differentiation of the small intestine. I. The accumulation of alkaline phosphomonoesterase in the duodenum of the chick. J. Exptl. Zool., **115**:109–129.

Ravn, E., 1896. Die Bildung des Septum Transversum beim Hühnerembryo. Arch. Anat. Entwg. 1896, pp. 157–186.

———, 1899. Ueber die Entwickelung des Septum Transversum. Anat. Anz., **15:** 528–534.

Schumacher, S., 1926. Die Entwicklung der Glandulae oesophageae des Hühnes. Jahrb. Morph. u. mikr. Anat., Abt. 2, **5:**11–22.

Seessel, A., 1877. Zur Entwickelungsgeschichte des Vorderdarms. Arch. Anat. Entwg., Jahrg. 1887: 449–467.

———, 1877. Zur Entwicklungsgeschichte des Vorderdarms. Arch. Anat. Physiol., Anat. Abt., pp. 449–467.

Simard, L. C., and Van Campenhout, E., 1932. The embryonic development of argentaffin cells in the chick intestine. Anat. Record, **53:**141–159.

Steding, G., and Klemeyer, L., 1969. Die Entwicklung der Perikardfalte des Huhnerembryo. Z. Anat. Entwickl.-gesch., **129:**223–233.

Van Alten, P. J., and Fennell, R. A., 1957. Histogenesis and histochemistry of the digestive tract of chick embryos. Anat. Record, **127:**677–695.

IX.   Urogenital System

Abel-Malek, E. T., 1950. Early development of the urinogenital system in the chick. J. Morphol., **86:**599–626.

Andres, G., 1955. Growth reactions of mesonephros and liver to intravascular injections of embryonic liver and kidney suspensions in the chick embryo. J. Exptl. Zool., **130:**221–249.

Atterbury, R. R., 1923. Development of the metanephric anlage of chick in allantoic grafts. Am. J. Anat., **31:**409–431.

von Berenberg-Gossler, H., 1912. Die Urgeschlechtszellen des Hühner-embryos am 3 and 4 Bebrütungstage mit besonderer Berücksichtigung der Kern-und Plasmastrukturen. Arch. f. mikr. Anat., 81, Abt. II, S. 24–72.

Boyden, E. A., 1922. The development of the cloaca in birds, with special reference to the origin of the bursa of Fabricius, the formation of a urodaeal sinus, and the regular occurrence of a cloacal fenestra. Am. J. Anat., **30:**163–201.

———, 1924. An experimental study of the development of the avian cloaca, with special reference to a mechanical factor in the growth of the allantois J. Exptl. Zool., **40:**437–471.

———, 1927. Experimental obstruction of the mesonephric ducts. Proc. Soc. Exptl. Biol. Med., **24:**572–576.

———, 1931. Correlation of uric acid production with growth of kidney tubules in chick embryos. Proc. Soc. Exptl. Biol. Med., **28:**625–626.

Chambers, R., and Kempton, R. T., 1933. Indications of function of the chick mesonephros in tissue culture with phenol red. J. Cellular Comp. Physiol., 3:131–167.

Ernst, M., 1926. Vergleichende Untersuchungen über die Urnierensekretion. Zeitschr. f. Anat. u. Entwg., 79:781–796.

Felix u. Bühler, 1906. Die Entwickelung der Harn-und-Geschlechtsorgane. Hertwig's Handbuch der vergleichenden und experimentellen Entwickelungslehre der Wirbeltiere, Fischer, Jena, III. Pt. 1, 81–442.

Gaarenstroom, J. H., 1939. Action of sex hormones on the development of the Müllerian duct of the chick embryo. J. Exptl. Zool., 82:31–46.

Goldsmith, J. B., 1935. The primordial germ cells of the chick. I. The effect on the gonad of complete and partial removal of the "germinal crescent" and of removal of other parts of the blastodisc. J. Morphol., 58:537–553.

Grafe, E., 1906. Beiträge zur Entwicklung der Urniere und ihrer Gefässe beim Hühnchen. Arch. f. mikr. Anat., 67:143–230.

Gruenwald, P., 1941. The relation of the growing Mullerian duct to the Wolffian duct and its importance for the genesis of malformations. Anat. Record, 81: 1–19.

——, 1942. Primary asymmetry of the growing Müllerian ducts in the chick embryo. J. Morphol., 71:299–305.

——, 1943. Stimulation of nephrogenic tissue by normal and abnormal inductors. Anat. Record, 86:321–339.

Juhn, M., and Gustavson, R. G., 1932. The response of a vestigial Müllerian duct to the female hormone and the persistence of such rudiments in the male fowl. Anat. Record, 52:299–311.

Kinyon, N., and Watterson, R. L., 1958. Reduced endocrine activity of ovaries of hypophysectomized duck embryos as indicated by modified development of genital tubercle and syrinx. Physiol. Zool., 31:60–72.

Lewis, L. B., 1946. A study of some effects of sex hormones upon the embryonic reproductive system of the white pekin duck. Physiol. Zool., 19:282–329.

Macdonald, E., and Taylor, L. W., 1933. The rudimentary copulatory organ of the domestic fowl. J. Morphol., 54:429–449.

Minoura, T., 1921. A study of testis and ovary grafts on the hen's egg and their effects on the embryos. J. Exptl. Zool., 33:1–61.

Mintz, B., and Wolff, E., 1954. The development of embryonic chick ovarian medulla and its feminizing action in intracoelomic grafts. J. Exptl. Zool., 126:511–536.

Nonidez, J. F., 1920. Studies on the gonads of the fowl. Hematopoietic processes in the gonads of embryos and mature birds. Am. J. Anat., 28:81–113.

Overton, J., 1959. Mitotic pattern in the chick pronephric duct. J. Embryol., Lond., 7:275–280.

Ramm, G. M., 1955. The fate of chick mesonephric tissue in intraocular homografts. J. Exptl. Zool., 130:107–132.

Reinhoff, W. F., 1922. Development and growth of the metanephros or permanent kidney in chick embryos (8–10 days incubation). Johns Hopkins Hosp. Bull., 33(381):392–406.

Swift, C. H., 1914. Origin and early history of the primordial germ-cell in the chick. Am. Jour. Anat., 15:483–516.

———, 1915. Origin of the definitive sex-cells in the female chick and their relation to the primordial germ-cells. Am. J. Anat., 18:441–470.

———, 1916. Origin of the sex-chords and definitive spermatogonia in the male chick. Am. J. Anat., 20:375–410.

Warner, F. J., 1927. Vestigial and provisional uriniferous tubules absent in the metanephros of birds. Anat. Record, 36:271–277.

Willier, B. H., 1927. The specificity of sex, of organization, and of differentiation of embryonic chick gonads as shown by grafting experiments. J. Exptl. Zool., 46:409–465.

Witshi, E., 1935. Origin of asymmetry in the reproductive system of birds. Am. J. Anat., 56:119–141.

Wolff, Etienne, and Haffen, K., 1952. Sur le developpement et la differenciation sexuelle des gonades embryonnaires d'oiseau en culture in vitro. J. Exptl. Zool., 119:381–404.

X.   Skeletal and Muscular Systems

Allen, M. J., 1948. The development of the brachial plexus and the wing-bud region of the chick embryo in chorio-allantoic grafts. J. Comp. Neurol., 89:41–69.

Amprino, R., and Camosso, M., 1955. Ricerche sperimentali sulla morfogenesi degli arti nel pollo. J. Exptl. Zool., 129:453–493.

Avery, G., Chow, M., and Holtzer, H., 1956. An experimental analysis of the development of the spinal column. V. Reactivity of chick somites. J. Exptl. Zool., 132:409–426.

Bevelander, G., Nakahara, H., and Rolle, G. K., 1960. The effect of tetracycline on the development of the skeletal system of the chick embryo. Develop. Biol., 2:298–312.

Blumstein-Judina, B., 1905. Die Pneumatisation des Markes der Vogelknochen. Anat. Hefte, Abt. I, 29:1–54.

Carpenter, E., 1950. Growth and form in vitro of avian femur rudiments on clots in which dried products are substituted for fresh fowl plasma. J. Exptl. Zool., 113:301–316.

Coulombre, A. J., and Crelin, E. S., 1958. The role of the developing eye in the morphogenesis of the avian skull. Am. J. Phys. Anthropol., 16:25–37.

Eastlick, H. L., 1943. Studies on transplanted embryonic limbs of the chick. I. The development of muscle in nerveless and in innervated grafts. J. Exptl. Zool., 93:27–49.

——— and Anderson, R. H., 1944. Studies on transplanted embryonic limbs of the chick. II. The development of limb primordia within the anterior chamber of the eye. J. Morphol., **75:**1–9.

———— and Wortham, R. A., 1947. Studies on transplanted embryonic limbs of the chick. III. The replacement of muscle by "adipose tissue." J. Morphol., **80:** 369–389.

Fell, H. B., 1925. The histogenesis of cartilage and bone in the long bones of the embryonic fowl. J. Morphol. and Physiol., **40:**417–459.

Hamburger, V., 1938. Morphogenetic and axial self-differentiation of transplanted limb primordia of 2-day chick embryos. J. Exptl. Zool., **77:**379–399.

Hooker, D., 1911. The development and function of voluntary and cardiac muscle in embryos without nerves. J. Exptl. Zool., **11:**159–187.

Isaacs, R., 1919. The structure and mechanics of developing connective tissue. Anat. Record, **17:**243–270.

Kingsbury, B. F., 1920. The developmental origin of the notochord. Science, N. S., **51:**190–193.

Lash, J., Holtzer, S., and Holtzer, H., 1957. An experimental analysis of the development of the spinal cord. VI. Aspects of cartilage induction. Exptl. Cell Res., **13:**292–303.

————, Holtzer, H., and Whitehouse, M. W., 1960. In vitro studies on chondrogenesis: the uptake of radioactive sulfate during cartilage induction. Develop. Biol., **2:**76–89.

Montagna, W., 1945. A re-investigation of the development of the wing of the fowl. J. Morphol., **76:**87–113.

O'Rahilly, R., and Gardner, E., 1956. The development of the knee joint of the chick and its correlation with embryonic staging. J. Morphol., **98:**49–88.

Paff, G. H., and Seifter, J., 1950. The effect of hyaluronidase on bone growth in vitro. Anat. Record, **106:**525–537.

Romer, A. S., 1927. The development of the thigh musculature of the chick. J. Morphol. Physiol., **43:**347–385.

Rudnick, D., 1945. Differentiation of prospective limb material from Creeper chick embryos in coelomic grafts. J. Exptl. Zool., **100:**1–17.

Saunders, J. W., and collaborators. See papers on development of appendages listed under IV. Early Differentiation of Embryo.

Sieglbauer, F., 1911. Zur Entwicklung der Vogelextremitat Zeitschr. f. wiss. Zool., **97:**262–313.

Spurling, R. G., 1923. The effect of extirpation of the posterior limb bud on the development of the limb and pelvic girdle in chick embryos. Anat. Record, **26:** 41–56.

Straus, W. L., and Rawles, M. E., 1953. An experimental study of the origin of the trunk musculature and ribs in the chick. Am. J. Anat., **92:**471–509.

Strudel, G., 1955. L'action morphogène du tube nerveux et de la corde sur la différenciation des vertèbres et des muscles vertébraux chez l'embryon de poulet. Arch. Anat. Microscop. Morphol. Exptl., **44:**209–235.

Sullivan, G. E., 1962. Anatomy and embryology of the wing musculature of the domestic fowl (Gallus). Australian J. Zool., **10:**458–518.

Warren, A. E., 1934. Experimental studies on the development of the wing in the embryo of Gallus domesticus. Am J. Anat., 54:449–485.

Watterson, R. L., Fowler, I., and Fowler, B. J., 1954. The role of the neural tube and notochord in development of the axial skeleton of the chick. Am. J. Anat., **95:**337–399.

Williams, L. W., 1910. The somites of the chick. Am. J. Anat., **11:**55–100.

Wortham, R. A., 1948. The development of muscles and tendons in the lower leg and foot of chick embryos. J. Morphol., **83:**105–148.

Zwilling, Edgar, 1955. Ectoderm-mesoderm relationship in the development of the chick embryo limb bud. J. Exptl. Zool., **128:**423–442.

————, 1956. Interaction between limb bud ectoderm and mesoderm in the chick embryo. I. J. Exptl. Zool., **132:**157–172; II. J. Exptl. Zool., **132:**173–188; III. J. Exptl. Zool., **132:**219–240; IV. J. Exptl. Zool., **132:**241–254.

XI. Pharyngeal Derivatives and the Ductless Glands

Adams, A. E., and Buss, J. M., 1952. The effect of a single injection of an antithyroid drug on hyperplasia in the thyroid of the chick embryo. Endocrinology, **50:**234–253.

Atwell, W. J., and Sitler, Ida, 1918. The early appearance of the anlagen of the pars tuberalis in the hypophysis of the chick. Anat. Record, **15:**181–187.

Bradway, W., 1929. The morphogenesis of the thyroid follicles of the chick. Anat. Record, **42:**157–167.

Brauer, A., 1932. A topographical and cytological study of the sympathetic nervous components of the suprarenal of the chick embryo. J. Morphol, **53:**277–325.

Carpenter, E., 1942. Differentiation of chick embryo thyroids in tissue culture. J. Exptl. Zool., **89:**407–431.

————, Beattie, J., and Chambers, R. D., 1954. The uptake of I^{131} by embryonic chick thyroid glands in vivo and in vitro. J. Exptl. Zool., **127:**249–270.

Dawson, A. B., 1953. Histochemical evidence of early differentiation of the suprarenal gland of the chick. J. Morphol., **92:**579–596.

Dossel, W. E., 1958. An experimental study of the locus of origin of the chick parathyroid. Anat. Record, **132:**555–562.

Douarin, N. L., et al., 1968. Étude des capacités de différenciation et du rôle morphogène de l'endoderme pharyngien chez l'embryon d'oiseau. Ann. Embryol. Morphogenese, **1:**29–39.

Dudley, J., 1942. The development of the ultimobranchial body of the fowl, Gallus domesticus. Am. J. Anat., **71:**65–97.

Friedman, B., 1934. The mesodermal relations of the pars buccalis of the hypophysis in the duck. J. Morphol., **55**:611–631.

Fugo, N. W., 1940. Effects of hypophysectomy in the chick embryo. J. Exptl. Zool., **85**:271–297.

Hamilton, H. L., and Hinsch, G. W., 1957. The fate of the second visceral pouch in the chick. Anat. Record, **129**:357–369.

Hammond, W. S., 1954. Origin of thymus in the chicken embryo. J. Morphol., **95**:501–522.

Hays, V. J., 1914. The development of the adrenal glands of birds. Anat. Record, **8**:451–474.

Hillemann, H. H., 1943. An experimental study of the development of the pituitary gland in chick embryos. J. Exptl. Zool., **93**:347–373.

Johnson, C. E., 1918. The origin of the ultimobranchial body and its relation to the fifth visceral pouch in birds. J. Morphol., **31**:583–597.

Kastschenko, N., 1887. Das Schlundspaltengebiet des Huhnchens. Arch. Anat. Entwg., Jg. 1887, pp. 258–300.

Mall, F. P., 1887. Entwickelung der Branchialbogen and -spalten des Huhnchens. Arch. Anat. Entwg., pp. 1–34.

Martindale, F. M., 1941. Initiation and early development of thyrotropic function in the incubating chick. Anat. Record, **79**:373–393.

Minervini, R., 1904. Des Capsules surrénales: Développement, structure, fonctions. J. de l'Anat. et de la Physiol., **40**:449–492; 634–667.

Okuda, M., 1928. The adrenalin content of the suprarenal capsule in the chick embryo. Endocrinology, **12**:342–348.

Poll, H., 1906. Die vergleichende Entwickelungsgeschichte der Nebennieren-systeme der Wirbeltiere. Herwig's Handbuch der vergleicheden und experimentellen Entwickelungslehre der Wirbeltiere, Fischer, Jena, III. Pt. 2, 443–618.

Rahn, H., 1939. The development of the chick pituitary with special reference to the cellular differentiation of the pars buccalis. J. Morphol., **64**:483–517.

Rudnick, D., 1932. Thyroid-forming potencies of the early chick blastoderm. J. Exptl. Zool., **62**:287–317.

Soulié, A. H., 1903. Recherches sur le développement des capsules surrénales chez les vertébrés supérieurs. J. de l'Anat. et de la Physiol., **39**:197–293.

Spiroff, B. E. N., 1958. Embryonic and post-hatching development of the pineal body of the domestic fowl. Am. J. Anat., **103**:375–401.

Sun, T. P., 1932. Histo-physiogenesis of the glands of internal secretion—thyroid, adrenal, parathyroid, and thymus of the chick. Physiol. Zool., **5**:384–396.

Willier, B. H., 1924. The endocrine glands and the development of the chick. I. The effects of thyroid grafts. Am. J. Anat., **33**:67–103.

———, 1930. Study of origin and differentiation of the suprarenal gland in the chick embryo by chorio-allantoic grafting. Physiol. Zool., **3**:201–225.

Wilson, M. E., 1952. The embryological and cytological basis of regional patterns in the definitive epithelial hypophysis of the chick. Am. J. Anat., **91**:1–50.

XII. *Experimental Methods and Tissue Culture Techniques*

(General treatises only. Experimental studies of the normal or abnormal development of specific structures are listed under the organ system concerned.)

Abercrombie, M., and Causey, G., 1950. Identification of transplanted tissues in chick embryos by marking with phosphorus-32. Nature, **166**:229–232.

Burrows, M. T., 1911. The growth of tissues of the chick embryo outside the animal body, with special reference to the nervous system. J. Exptl. Zool., **10**:63–83.

DeHaan, R. L., 1958. Cell migration and morphogenetic movements. Pp. 339–374 in A Symposium on the Chemical Bases of Development. (Edited by Elroy and Glass.) The Johns Hopkins Press, Baltimore.

Dent, J. N., and Hunt, E. L., 1954. Radiotracer techniques in embryological research. J. Cellular. Comp. Physiol., **43**(suppl. 1):77–102.

Fischer, A., 1927. Gewebezuchtung. Handbuch der Biologie der Gewebezellen in Vitro. Muller and Steinicke, Munchen. xv + 508 pp.

Hamburger, V., 1942. A Manual of Experimental Embryology. University of Chicago Press. 204 pp.

Harkmark, W., and Graham, K., 1951. A method for opening and closing eggs with an experimental demonstration of its advantages. Anat. Record, **110**:41–48.

Harrison, J. R., 1954. Growth and differentiation of the embryonic chick eye in vitro. I. On yolk-albumen fractions obtained with centrifugation and heat. J. Exptl. Zool., **127**:493–510.

Hillemann, H. H., 1942a. A method and apparatus for opening and closing eggs which will permit turning at regular intervals during subsequent incubation. Anat. Record, **84**:331–336.

————, 1942b. An embryological micromanipulator with an adjustable egg-holder attachment. Anat. Record, **84**:337–342.

————, 1942c. The design and use of micro-electrodes for the production of lesions in the pituitary rudiment of chick embryos. Anat. Record, **84**:343–357.

Huxley, J. S., and DeBeer, G. R., 1934. The Elements of Experimental Embryology. Cambridge University Press, London. xiii + 514 pp.

Jordanov, J., 1960. Cultivation of chick embryo tissues in egg yolk dialysates. Acta biol. med. Germanica, **4**:233–246.

Kutsky, R. J., and Harris, M., 1954. Growth promoting agents in cell fractions of chick embryos. J. Cellular Comp. Physiol., **43**:193–204.

Lewis, M. R., 1917. Development of connective-tissue fibers in tissue cultures of chick embryos. Carnegie Cont. to Embryol., No. 17, **6**:45–60.

———— and Lewis, W. H., 1911. The cultivation of tissues from chick embryos in solutions of NaCl, CaCl$_2$, KCl, and NaHCO$_6$. Anat. Record, **5**:277–293.

Lewis, W. H., 1923. Amniotic ectoderm in tissue cultures. Anat. Record, **26**:97–117.

————, 1926. Cultivation of embryonic heart-muscle. Carnegie Cont. to Embryol., No. 90, **18**:1–21.

———— and Lewis, M. R., 1912a. The cultivation of sympathetic nerves from the intestine of chick embryos in saline solutions. Anat. Record, **6**:7–32.

———— and Lewis, M. R., 1912b. The cultivation of chick tissues in media of known chemical constitution. Anat. Record, **6**:207–211.

McWhorter, J. E., and Whipple, A. C., 1912. The development of the blastoderm of the chick in vitro. Anat. Record, **6**:121–140.

New, D. A. T., 1966. The Culture of Vertebrate Embryos. Academic Press Inc., New York. xi + 245 pp.

Parker, R. C., 1961. Methods of Tissue Culture. Paul B. Hoeber, New York. 3d ed., 358 pp.

Romanoff, A. L., 1931. Cultivation of the chick embryo in an opened egg. Anat. Record, **48**:185–189.

Rugh, R., 1962. Experimental Embryology. Burgess Publishing Company, Minneapolis. 3d ed., ix + 501 pp.

Schneider, H., Shaw, M. W., Muirhead, E. E., and Smith, A., 1965. The *in vitro* culture of embryonic chicken heart cells. Exptl. Cell Res., **39**:631–636.

Spratt, N. T., Jr., 1947a. A simple method for explanting and cultivating early chick embryos *in vitro*. Science, **106**:452.

————, 1947b. Development in vitro of the early chick blastoderm explanted on yolk and albumen extract saline-agar substrata. J. Exptl. Zool., **106**:345–365.

————, 1948. Development of the early chick blastoderm on synthetic media. J. Exptl. Zool., **107**:39–64.

————, 1949. Nutritional requirements of the early chick embryo. I. The utilization of carbohydrate substrates. J. Exptl. Zool., **110**:273–298.

Taylor, K. M., and Schechtman, A. M., 1949. *In vitro* development of the early chick embryo in the absence of small organic molecules. J. Exptl. Zool., **111**:227–253.

Vogelaar, J. P. M., and van den Boogert, J. B., 1925. Development of the egg of Gallus domesticus in vitro. Anat. Record, **30**:385–395.

Waddington, C. H., 1932. Experiments on the development of chick and duck embryos, cultivated *in vitro*. Phil. Trans. Roy. Soc. London, Ser. B, **221**:179–230.

Weiss, P., 1939. Principles of Development. A Text in Experimental Embryology. Holt, Rinehart and Winston, Inc., New York. xix + 601 pp.

Willier, B. H., Weiss, P. A., and Hamburger, V., 1955. Analysis of Development. W. B. Saunders Company, Philadelphia. xii + 735 pp.

XIII.   Teratogens and Other Means of Disturbing Normal Development

Adams, A. E., 1958. Effects of thyroxine on chick embryos. Anat. Record, **131**:445–463.

———— and Bull, A. L., 1949. The effects of antithyroid drugs on chick embryos. Anat. Record, 104:421–443.

Adamstone, F. B., 1931. The effects of vitamin-E deficiency on the development of the chick. J. Morphol. Physiol., 52:47–89.

Alsop, F. M., 1919. The effect of abnormal temperatures upon the developing nervous system in chick embryos. Anat. Record, 15:307–331.

Beams, H. W., and King, R. L., 1936. The effect of ultracentrifuging upon chick embryonic cells, with special reference to the "resting" nucleus and the mitotic spindle. Biol. Bull., 71:188–198.

Beaudoin, A. R., 1961. Teratogenic activity of several closely related disazo dyes on the developing chick embryo. J. Embryol. Exptl. Morphol., 9:14–21.

Bless, A. A., and Romanoff, A. L., 1943. Stimulation of the development of the avian embryo by x-rays. J. Cellular Comp. Physiol., 21:117:121.

Buchanan, J. W., 1926. Regional differences in rate of oxidations in the chick blastoderm as shown by susceptibility to hydrocyanic acid. J. Exptl. Zool., 45:141–157.

Bueker, E. D., and Platner, W. S., 1956. Effect of cholinergic drugs on development of chick embryo. Proc. Soc. Exptl. Biol. Med., 91:539–543.

Byerly, T. C., 1926a. Studies in growth. I. Suffocation effects in the chick embryo. Anat. Record, 32:249–270.

————, 1926b. Studies in growth. II. Local growth in "dead" chick embryos. Anat. Record, 33:319–325.

Catizone, O., and Gray, P., 1921. Experiments on chemical interference with the early morphogenesis of the chick. II. The effects of lead on the central nervous system. J. Exptl. Zool., 87:71–83.

Chapman, A. O., 1954. The effects of varying dosage of radioactive phosphorus (P^{32}) in the early chick embryo. J. Morphol., 95:451–470.

Colwell, H. A., Gladstone, R. J., and Wakely, C. P. G., 1926. Action of x-rays upon the developing chick embryo. Series 2 and 3. J. Anat., 60:207–228.

Corliss, C. E., Fedinec, A. A., and Robertson, G. G., 1966. The teratogenic effect of tetanus toxin on the central nervous system of the early chick embryo. Anat. Record, 154:221–232.

Couch, J. R., Cravens, W. W., Elvehjem, C. A., and Halpin, J. G., 1948. Relation of biotin to congenital deformities in the chick. Anat. Record, 100:29–48.

Cravens, W. W., McGibbon, W. H., and Sebesta, E. E., 1944. Effect of biotin deficiency on embryonic development in the domestic fowl. Anat. Record, 90:55–64.

Dagg, C. P., and Karnofsky, D. A., 1955. Teratogenic effects of azaserine on the chick embryo. J. Exptl. Zool., 130:555–572.

Dongen, R., van, 1964. Insulin and myeloschisis in the chick embryo. Australian J. Exptl. Biol. Med. Sci., 42:607–614.

Douarin, G. L., and Kirrmann, J., 1964. Action des rayons X sur la teneur en coen-

zyme I et sur l'activité de la coenzyme I- glycohydrolase chez le coeur embryonnaire du poulet en culture organotypique. Compt. Rend. Acad. Sci. Paris,
259:2153–2155.

Ebert, J. D., 1950. An analysis of the effects of anti-organ sera on the development, in vitro, of the early chick blastoderm. J. Exptl. Zool., **115**:351–378.

———— and Wilt, F. H., 1960. Animal viruses and embryos. Quart. Rev. Biol., **35**:
261–312.

Ehmann, B., 1964. The effect of thalidomide on chick embryos. Klin. Wochschr.,
42:295–296.

Gabriel, M. L., 1946a. The effect of local applications of colchicine on Leghorn and
polydactylous chick embryos. J. Exptl. Zool., **101**:339–350.

————, 1946b. Production of strophosomy in the chick embryo by local applications of colchicine. J. Exptl. Zool., **101**:351–354.

Grabowski, C. T., 1961a. Lactic acid accumulation as a cause of hypoxia-induced
malformations in the chick embryo. Science, **134**:1359–1360.

————, 1961b. A quantitive study of the lethal and teratogenic effects of hypoxia
on the three-day chick embryo. Am. J. Anat., **109**:25–35.

————, 1966a. Teratogenic effects of calcium salts on chick embryos. J. Embryol.
Exptl. Morph., **15**:113–118.

————, 1966b. Physiological changes in the bloodstream of chick embryos exposed
to teratogenic doses of hypoxia. Develop. Biol., **13**:199–213.

———— and Parr, J. A., 1958. The teratogenic effects of graded doses of hypoxia on
the chick embryo. Am. J. Anat., **103**:313–347.

———— and Schroeder, R. E., 1968. A time-lapse photographic study of chick embryos exposed to teratogenic doses of hypoxia. J. Embryol. Exptl. Morphol.,
19:347–362.

Hammett, F. S., and Wallace, F. L., 1928. Biology of metals. VII. Influence of lead
on the development of the chick embryo. J. Exptl. Med., **48**:659–665.

Herrmann, H., Donigsberg, U. R., and Curry, M. F., 1955. A comparison of the effects of antagonists of leucine and methionine on the chick embryo. J. Exptl.
Zool., **128**:359–377.

Hessel, D. L., and Stearner, S. P., 1964. Some effects of Co^{60} gamma radiation on
chick embryos. Argonne Nat. Lab. Biol. Med. Res. Div. Semiannual Rept.,
5–14.

Hinrichs, M. A., 1927. Modification of development on the basis of differential susceptibility to radiation. 4. Chick embryos and ultraviolet radiation. J. Exptl.
Zool., **47**:309–342.

Källén, B., and Rudeberg, S., 1964. Teratogenic effects of electric light on early chick
embryos. Acta Morphol. Neerl. Scand., **6**:95–99.

Karnofsky, D. A., and Lacon, C. R., 1961. Effects of physiological purines on the
development of the chick embryo. Biochem. Pharmacol., **7**:154–158.

Kirrmann, J. M., 1967. Radiobiologie - Effects des rayons X sur les acides nucleiques et l'incorporation d'uridine et de thymidine tritiées dans un intestin embryonnaire de poulet cultivé in vitro. Compt. Rend. Acad. Sci. Paris, **265:**512–515.

Landauer, W., 1929. Experimental studies concerning the development of the chicken embryo. I. The toxic action of lithium and magnesium salts. Poultry Sci. **8:** 301–312.

―――, 1936. Micromelia of chicken embryos and newly hatched chicks caused by a nutritional deficiency. Anat. Record, **64:**267–276.

―――, 1953. On teratogenic effects of pilocarpine in chick development. J. Exptl. Zool., **122:**469–484.

―――, 1956. The teratogenic activity of pilocarpine, pilocarpidine and their isomers, with special reference to the importance of steric configuration. J. Exptl. Zool., **132:**39–50.

Lutz, H., Bonhomme, C., and Lutz-Ostertag, Y., 1955. Action des ultrasons sur le blastoderme non incude d'Oiseau. Compt. Rend. Séances Acad. Sci., **240:** 1931–1932.

Lutz-Ostertag, Y., and Sénaud, J., 1966. Action de l'extrait de toxoplasme (toxoplasmine) sur le développement embryonnaire du poulet. Arch. Anat. Micro. Morphol. Expt., **55:**363–385.

Miclea, C., and Arcan, A., 1966. Effects of intraependymal injections with Janus green B and tretamine upon chick embryo development. Rev. Roumaine Embryol. Cytol., **3:**75–85.

Moscona, M. H., and Darnofsky, D. A., 1960. Cortisone-induced modifications in the development of the chick embryo. Endocrinology, **66:**533–549.

Nelsen, O. E., 1955. The effects of increased atmospheric pressures upon early chick development. J. Morphol., **96:**359–380.

Overton, J., 1958. Effects of colchicine on the early chick blastoderm. J. Exptl. Zool., **139:**329–347.

Paff, G. H., 1939. The action of colchicine upon the 48-hour chick embryo. Am. J. Anat., **64:**331–349.

Pierro, L. J., 1961. Teratogenic action of actinomycin D in the embryonic chick. II. Early development. J. Exptl. Zool., **148:**241–249.

Richter, K. M., 1956. Studies on the individual and joint effects of histamine and an antihistamine on growth, contractility and plasmocrine activity in cultures of embryonic chick heart. J. Cellular Comp. Physiol., **48:**147–166.

Robertson, G. G., Williamson, A. P. and Blattner, R. J., 1955. A study of abnormalities in early chick embryos inoculated with Newcastle disease virus. J. Exptl. Zool., **129:**5–44.

Romanoff, A. L., and Bauernfeind, J. C., 1942. Influence of riboflavin-deficiency in eggs on embryonic development (Gallus domesticus). Anat. Record, **82:**11–23.

Rous, P., and Murphy, J. B., 1911. Tumor implantations in the developing embryo. J. Am. Med. Assoc., **56:**741–742

Salzgeber, B., 1964. The genesis of phocomelia: limb-bud alterations in chickens treated with "nitrated yperite." Compt. Rend. Acad. Sci., Paris, **259**:2149–2151.

Sedar, J. D., 1956. The influence of direct current fields upon the developmental pattern of the chick embryo. J. Exptl. Zool., **133**:47–72.

Seichert, V., and Jelinek, R., 1967. Quantitative effect of the trypan blue in growing chicken embryo. Rev. Roumaine Embryol. Cytol., **4**:7–21.

Stiles, K. A., and Watterson, R. L., 1937. The effects of jarring upon the embryogeny of chick embryos. Anat. Record, **70**:7–12.

Stillwell, E. F., 1952. The effect of elevated temperature upon the occurrence of multipolar mitosis in embryonic cells grown in vitro. Anat. Record, **112**:195–215.

Woodard, T. M., Jr., and Estes, S. B., 1944. Effect of colchicine on mitosis in the neural tube of the forty-eight hour chick embryo. Anat. Record, **90**:51–54.

Zwilling, E., and DeBell, J. T., 1950. Micromelia and growth retardation as independent effects of sulfanilamide in chick embryos. J. Exptl. Zool., **115**:59–81.

XIV. General Distortions; Double Embryos

Andres, G., 1953. Experiments on the fate of dissociated embryonic cells (chick) disseminated by the vascular route. Part II. Teratomas. J. Exptl. Zool., **122**: 507–540.

Bollinger, O. P., 1928. Twinning in chick embryos. Proc. Pennsylvania Acad. Sci., **2**:106–107.

Fortuyn-van Leyden, C. E. Droogleever, 1927. Triple monstrum of the embryo of a chicken. Anat. Record, **34**:233–236.

Franke, K. W., Moxon, A. L., Poley, W. E., and Tully, W. C., 1936. Monstrosities produced by the injection of selenium salts into hens' eggs. Anat. Record, **65**:15–22.

Grabowski, C. T., 1957. The induction of secondary embryos in the early chick blastoderm by grafts of Hensen's node. Am. J. Anat., **101**:101–134.

Hutt, F. B., and Greenwood, A. W., 1929. Studies in embryonic mortality in the fowl. 3. Chick monsters in relation to embryonic mortality. Proc. Roy. Soc. Edinburgh, **49**:145–155.

Lutz, H., 1949. Sur la production expérimentale de la polyembryonie et de la monstruosité double chez les oiseaux. Arch. Anat. Micr. Morphol. Expt. **38**:79–114.

———, 1953. La duplication des organes axiaux d'embryons d'Oiseau après la formation de la ligne primitive. Compt. Rend Séances Acad. Sci., **236**:1825–1827.

O'Donoghue, C. H., 1910. Three examples of duplicity in chick embryos with a case of ovum in ovo. Anat. Anz., **37**:530–536.

Riddle, O., 1923. On the cause of twinning and abnormal development in birds. Am. J. Anat., **32**:199–252.

Roberts, E., Card, L. E., and Boyden, E. A., 1929. Double vents, associated with congenital absence of the left kidney and persistence of the embryonic genital papilla, in the domestic fowl. Anat. Record, **43**:155–163.

Schwalbe, E., 1906–1913. Die Morphologie der Missbilbungen des Menschen und ter Tiere. Fischer, Jena.

Stockard, C. R., 1921. Developmental rate and structural expression: An experimental study of twins, "double monsters," and single deformities, and the interaction among embryonic organs during their origin and development. Am. J. Anat., **28**:115–277.

Tannreuther, G. W., 1919. Partial and complete duplicity in chick embryos. Anat. Record, **16**:355–367.

Wolff, E., 1933. Recherches sur la structure d'omphalocéphales obtenus expérimentalement (mémoire préliminaire). Arch. Anat. Histol. Embryol., **16**:135–193.

———, 1934. Recherches expérimentales sur l'otocéphalie et les malformations fondamentales de la face (mémoire préliminare). Arch Anat. Histol. Embryol., **18**:229–262.

———, 1948. La duplication de l'axe embryonnaire et la polyembryonie chez les Vertébrés. Compt. Rend. Soc. Biol., **142**:1282–1306.

———and Lutz, H., 1947. Sur une méthode permettant d'obtenir expérimentalement le dédoublement des embryons d'oiseaux. Compt. Rend. Soc. Biol., **141**:901–903.

Wright, R., 1906. An early Anadidymus of the chick. Trans. Roy. Soc. Canada, **12**:21–26.

INDEX

Where feasible like structures are grouped together. For example, look up "Artery, coeliac" rather than "Coeliac artery." References in **boldface** refer to pages on which the subject indexed appears in an illustration; there may or may not be relevant text material on the same page. The color illustrations are tipped-in as unnumbered pages. They are referred to as opposite (*op.* 27) the page where they are inserted.